Psychotronics
and a Biodynamic Garden

*The author mows a cover crop at the
Kerr Center for Sustainable Agriculture, Poteau, Oklahoma*

*How to Grow and Harvest
Healthier Food through
Radionics and Dowsing*

Psychotronics
and a Biodynamic Garden

George Kuepper

Portal books 2021

Portalbooks

An imprint of SteinerBooks / Anthroposophic Press
PO Box 58, Hudson, NY 12534
www.steinerbooks.org

Copyright © 2021 by George Kuepper
and SteinerBooks / Anthroposophic Press, Inc.

All rights reserved. No part of this publication may be reproduced, stored in a retrieval system, or transmitted, in any form or by any means, electronic, mechanical, photocopying, recording, or otherwise, without the prior written permission of the publisher.

COVER DESIGN: Mary Giddens
COVER PHOTO: The author with a beneficial habitat planting of sunflower and sesame at the Cannon Horticulture Site, Poteau, Oklahoma (by Kerr Center for Sustainable Agriculture, Poteau, Oklahoma
BOOK DESIGN: Jens Jensen

The author's website:
www.midsouthradionics.com

LIBRARY OF CONGRESS CONTROL NUMBER: 2021931959

ISBN: 978-1-938685-31-6 (paperback)
ISBN: 978-1-938685-32-3 (eBook)

Contents

	Preface: How to Understand this Book	vii
	Introduction	xi
1.	Psychotronics	1
2.	How to Dowse	37
3.	Establishing Projects and Priorities	53
4.	Plant–Earth Alignment	57
5.	Intent and Manifestation	62
6.	Psychotronics Broadcasting	71
7.	More about Witnesses and Protocols	84
8.	Recycling Frequencies: An Application of Psychotronics	95
9.	Rationale for the Plant-positive Approach	99
10.	Biodynamics	106
11.	Nature Beings, Psychotronics, and Co-creative Gardening	133
12.	Balancing the Whole Farm and Homestead	137
13.	Balancing the Garden and Field	152
14.	Balancing Individual Crops and Plants	171
	Appendices:	
	Appendix 1: The Evolution of Radionics and Psychotronics for Farming and Gardening	183
	Appendix 2: Resources	203
	Cited Works	205
	Related Reading	210
	About the Author	213

for
Deede

Preface:
How to Understand this Book

Back in the 1990s, I wrote two books: *Radionics, Reality, and Man* and *Plants, Soils, Earth Energy, and Radionics.* It had been only a decade since my first psychotronics class, and a few experienced practitioners suggested I was a bit presumptive to publish anything on this subject. Perhaps that was so, but the prominent researchers, practitioners, and educators of that time were doing little to provide the how-to training manuals that novices needed. So, even though my knowledge and experience were limited and my writing style pedestrian, I went ahead and did what I did. I'm not apologetic.

A bit more than twenty years have sailed by and I've written a third book. This one, like *Plants, Soils, Earth Energy, and Radionics,* deals with growing plants and food. I've attempted to marry psychotronics with biodynamics (BD)—a spiritually based approach to gardening and farming that I'm growing to love and appreciate more and more. Though I'm now fairly well-seasoned in psychotronics, I'm quite the novice to biodynamics. So, in the eyes of experienced BD scholars and practitioners, I am, once again, an upstart. They're probably right.

But my dilemma is the same. With the exception of the writings of Hugh Lovel (his fine books and countless articles in *Acres U.S.A.* magazine), little has been written about merging psychotronics and biodynamics. Once again, I feel the urge to pull together the dribs and drabs of things I've learned from others and from my own practical experience.

Psychotronics and a Biodynamic Garden

Please don't misunderstand. I do not possess unique access to the mysteries of nature. I'm not clairvoyant. Nor do I claim much original thinking. Largely everything I write about originated with someone else and builds on their work. This is a compelling reason for the large number of references I use throughout the book. Another reason is my desire to preserve and pass on information. Good references are a trail of bread crumbs that link future researchers and practitioners back to the sources of truly original thinking.

Two Important Ideas

One of the most important ideas I hope the reader will capture is the value and flexibility of dowsing. If you learn to dowse, either through this book or through some other means, you are likely to find it the most empowering skill you have—one that is literally at your fingertips. I dowse so very many times every day, for so many reasons, and wouldn't want it any other way. But to be clear, I am neither frivolous nor capricious about dowsing. I do not dowse for information I don't intend to act on. If my only need is to satisfy my curiosity, I use the internet or the library.

The second important matter is the concept of intelligent beings in nature—the devas and nature spirits that I will later refer to as co-creative partners. I wrestled with my cautious self over addressing this subject; I fear it might be a "bridge too far" for many readers. But had I left it out, I'd have stripped the book of its most compelling content. Still, if you find such notions off-putting, you will find that most of the psychotronic and biodynamic techniques I detail can be performed without consciously acknowledging spiritual beings in nature. Personally, though, I think you'll lose something of value.

Don't Try to Do It All

I do not intend this book as a roadmap for others, but it *is* the record of my journey—the roadmap I've followed while exploring how psychotronics and biodynamics might be joined in the pursuit of a more sustainable and spiritually conscious agriculture. I hope, however, that readers discover ideas and methods that will be interesting and useful to them. Should they find something that truly helps them to grow food or enhance the natural world, that would give me great joy.

G. L. Kuepper
Spring 2020
Fayetteville, Arkansas

The author with an early-fall harvest
of Creek Indian Pumpkin for seed propagation.
(photo taken adjacent to his biodynamic garden near Goshen, Arkansas)

Introduction

I grew up on a small, southeastern Wisconsin dairy farm in the 1950s and 1960s. I loved that life, but did not see agriculture in my future. At age fourteen, I headed off to Catholic preparatory seminary, thinking I might have a "calling" to the priesthood. Apparently not. After four years of an excellent education and moral guidance by the Capuchins, I found myself at the University of Wisconsin, Madison.

By then, I'd missed the Summer of Love in '67 and Woodstock in '69, but I was in plenty of time for the social and political turbulence of the early and mid-'70s, including the first Earth Day, pioneered by Wisconsin's own Senator Gaylord Nelson in April, 1970.

I can't claim I made a beeline back to agriculture. Frankly, I was disappointed by the unquestioning commitment of the University's faculty to using pesticides and other chemicals with little or no concern for the environment or food quality. This led me to explore other paths for a while but, eventually, I returned to the University of Wisconsin to earn a Master's Degree in agronomy in 1977.

I had several interesting prospects on graduation but chose the one that paid the least—a research assistant position at the Center for the Biology of Natural Systems at Washington University in St. Louis, founded by Dr. Barry Commoner (1917–2012). I jumped at this job for a single reason. One of Commoner's research teams, led by Willie Lockeretz (1943–2019), was conducting the first in-depth evaluation of organic agriculture in the United States.

For roughly three years, I visited and studied dozens of commercial organic farms and farmers in the western Corn Belt. I helped collect and evaluate soil samples, crop yields, and ecological and economic data. They unveiled an approach to agriculture that was easier on the environment, consumed fewer resources, and was economically competitive with similar conventional farms, even when selling into the same conventional marketplace.

It was during these years that I first heard of biodynamics. I was intrigued, but couldn't understand or relate to it. I tried to read Rudolf Steiner's Agriculture Course several times, but never got far. I shelved biodynamics, along with other worthy but seemingly incomprehensible ideas, in the back recesses of my mind for some future time.

In the mid-1980s, my wife and I moved to her home state of Oklahoma. There I joined the Kerr Foundation's Agriculture Division, headquartered near Poteau. Little did I know that the Ag Division would, within months, reinvent itself as the Kerr Center for Sustainable Agriculture.

My first major task was to develop a diverse pick-your-own demonstration farm designed around a model made popular by Dr. Booker T. Whatley (1915–2005).[1] We began the project by purchasing a young three-acre blueberry patch about five miles from the Foundation's headquarters. The blueberry planting, itself, was a huge challenge. Though well-intentioned, the previous grower had misused and overused conventional fertilizers in his effort to compensate for a marginal location. The soil was imbalanced and the plants were stressed; they drew in virtually every blueberry pest and disease known to the region plus a plethora of weeds. The grower resorted to herbicides, insecticides, and fungicides, often making the situation worse.

At the time I became involved, the costs of weed and pest control, plus annual plant replacement, were already unsustainable. Seeking guidance, I tried a range of university testing and

---

[1] Booker T. Whatley, "The Small Farm Plan," *Mother Earth News*. May/June 1982 (https://www.motherearthnews.com/homesteading-and-livestock/small-farm-plan-zmaz82mjzkin).

consultants. They did not help much. Relying on my organic knowledge, I began using more benign inputs, but these seemed too little and too late. Blueberries were the linchpin of our farm model so the entire project was in jeopardy. It was then that I stumbled onto radionics.

I took my first radionics–psychotronics training at a sugarcane plantation near Mercedes, Texas. I was a poor student—slow to understand and not quick to pick up essential skills. Perhaps I was trying too hard; maybe I put myself under too much pressure.

Despite my dismal class performance, I returned to the farm and attempted to apply what I'd learned to the failing blueberries. I focused on radionics analyses and started to formulate foliar nutrition sprays. This was one of the techniques I was most drawn to during training and felt I could do successfully. I hoped that foliar feeding (radionically guided) might support the crop and encourage soil and plant root systems to heal themselves.

To my great satisfaction, the strategy worked. The changes appeared almost immediately. All insect pests and diseases retreated; some vanished completely. The weed populations changed from serious problems, such as Bermuda grass, to much less challenging annual species. Most importantly, the entire planting became healthy and vigorous. Yields rose to competitive economic levels and we never had to replace another plant. I soon began applying these techniques to other crops with comparable success.

As time went on, I occasionally revisited biodynamics. I understood that it shared a similar world view with psychotronics. I was especially influenced by the writings of Hugh Lovel (1947–2020), who artfully brought psychotronics and biodynamics together in his own way. However, it was not until I was semi-retired from the Kerr Center that I would attempt to merge the two in my own fashion. This book presents the background to that effort and my progress to the present.

The author dowses with an SE-5 radionics instrument

1

Psychotronics

One of the most useful lessons learned from writing my first book on radionics in the mid-1990s was how it increased my knowledge and understanding of the topic. Once you seriously undertake the challenge of explaining a subject to others, it quickly exposes the gaps, contradictions, and fallacies in what you thought you knew. That happened often while writing this book. And despite close to thirty-four years of experience with psychotronics, it certainly humbled me as I researched and wrote this particular chapter.

Psychotronics covers a range of modalities (including dowsing and radionics) that can be used to access and study the hidden reality behind our physical world. It provides us with practical means for investigating, navigating, and even changing this reality. Without a doubt, psychotronics is controversial and the amount of misinformation and disinformation surrounding it is dizzying. So, understand that what I'm presenting here is a *working paradigm*. It will explain how *I* have come to understand psychotronics and how *I* am using it.

What Is a Working Paradigm?

Paradigms are concepts, thought patterns, theories, and standards that help us make sense of the world. Our cultural, political, religious, scientific, and personal paradigms guide us when we choose what to accept as real or not. While paradigms do not exclude new information and ideas, they consciously and subconsciously set rules for what can and cannot become part of our reality.

Paradigms are often rife with contradictions. America, today, is heavily influenced by paradigms of Christianity *and* materialism. These two paradigms often conflict, but most of us seem to handle it without having a deep personal crisis. So, while materialism and physical science do not readily accommodate spiritual, theological, metaphysical, or magical concepts, there are many church-going scientists and materialists who are satisfied with some form of dualistic reality in which both the material and nonmaterial spiritual realms coexist. Perhaps this is a matter of compartmentalization, where the relationship between compartments is weakly rationalized or is simply not examined.

In this chapter I am presenting what I call my *working paradigm*. It is a paradigm in which psychotronics and biodynamics make sense to me. To the best of my ability, I also try to explain *why* they make sense to me. This working paradigm recognizes both the physical and the nonphysical. But unlike dualistic models, mine does not compartmentalize; rather, it is built on the notion that there is a dynamic seamless relationship between material and spiritual realms.

I've tried to explain my working paradigm using both modern scientific *and* traditional spiritual concepts, but there are a lot of these concepts I continue to struggle with and might not interpret them as well as I should. So, you might need to cut me some slack here and there.

Parapsychology

I'm beginning by defining the broad term—*parapsychology*. Parapsychology is the study of paranormal and psychic phenomena. It is an umbrella term that covers subjects like telepathy, precognition, clairvoyance, psychokinesis, near-death experiences, reincarnation, apparitional experiences, and other paranormal studies. It also covers the topic we address here: psychotronics.

What Is Psychotronics?

The United States Psychotronics Association (USPA) defines *psychotronics* as "the science of mind–body–environment relationships, an interdisciplinary science concerned with the interactions of matter, energy, and consciousness."[1] The term *psychotronics* emerged from parapsychology studies done in Czechoslovakia during the 1960s, which involved the mind (*psycho*) operating devices (*tron*) using thought. Early psychotronics researchers conceived of and built numerous psychotronic devices. Often, these were sacred geometrical forms composed of pyramids and cones along with crystals, copper coils, and other materials. These devices were (and remain) dependent on a human operator. As a result, their performance varied depending on the specific practitioner involved. Though somewhat unpredictable, they produced, and continue to produce, positive effects on people, animals, plants, and natural processes.[2]

While psychotronics is included under the umbrella term *parapsychology*, *psychotronics* is, in turn, an umbrella term that encompasses dowsing and radionics.

Defining Dowsing

Dowsing is covered by the umbrella of psychotronics, because it entails using the mind to operate devices. The devices, in this case, are simple ones—including forked sticks, pendulums, bobbers, and L-rods.

A simple and straightforward definition of *dowsing* comes from Christopher Bird (1928–1996), the author of the 340-page compendium, *The Divining Hand*: "To dowse is to search with the aid of a

---

1 US Psychotronics Association (USPA): www.psychotronics.org/about.
2 Rubik, Beverly, "What Is Psychotronics?" US Psychotronics Association: https://www.psychotronics.org/about.

handheld instrument such as a forked stick or a pendular bob on the end of a string—for anything..."[3]

Dowsing has many other names—divining, doodling, witching, rhabdomancy, and radiesthesia, among others. We sometimes use these terms interchangeably. More often, though, we associate each with a specific application or technique of dowsing. *Radiesthesia*, for example, is most commonly associated with dowsing for human health issues; *rhabdomancy* implicitly indicates that sticks or wands are used.[4]

The Origins of Dowsing

Consensus holds that dowsing has prehistoric origins. This belief is based largely on interpretations of ancient stories and drawings. One source, for example, cites:

- a 9,000-year-old cave painting in Africa that appears to depict a dowser in action;
- a 9,000-year-old rock carving in Peru of a man holding a forked stick;
- a 4,000-year-old statue of Chinese Emperor Kwang Yu holding a forked stick;
- a 2,300-year-old painting from the Middle East featuring a priest appearing to dowse.[5]

Bird dedicates a whole chapter in *The Divining Hand* to rhabdomancy—dowsing using a wand or stick.[6] He describes croziers, scepters, and similar rods that historically represented power and, in many instances, the ability to *divine*.[7] A subsequent chapter

---

[3] Bird, *The Divining Hand: The 500-Year-Old Mystery of Dowsing*, p. 1 (for this and other cited works, please see bibliography for details).

[4] Ibid., p. 63.

[5] Yarrow, *Dowsing History and Techniques: Beginners Guide*, p. 3.

[6] Bird, *The Divining Hand*, pp. 59-76.

[7] *Divining* has been defined as discovering something by intuition or insight (*Merriam-Webster*. 2020).

moves on to more recent historical record. The author provides documentation that dowsing was in common use from the 1500s onward to help site mines and water wells.[8]

Robert Leftwich writes something particularly thoughtful about the origins of dowsing in his little book, *Dowsing: The Ancient Art of Rhabdomancy*: "Because the so-called gift of dowsing is in reality one of our birthrights, it has always existed and consequently its origin is lost in antiquity."[9] This quotation highlights one of the core beliefs about dowsing—that our bodies are naturally endowed with the ability to dowse, and that, whereas a few individuals are exceptionally gifted in the craft, virtually all of us are capable of doing it to some useful degree.

Dowsing and Intuition

Dowsing exercises a human faculty that allows us to obtain information in a manner beyond the scope and power of our standard physical senses.[10] It is the same faculty that undergirds human intuition or "gut feeling"—our so-called *sixth sense*. We usually think of intuition as feelings that come to us without advance pondering or thought. These feelings suddenly appear and can mysteriously link us with relevant and often important information.[11]

Although they involve the same faculty, intuition and dowsing have differences. The most obvious is that intuition does not require dowsing tools like rods of pendulums. But furthermore, intuition is a sudden and direct perception of some truth or fact without *reasoning*. Dowsing, on the other hand, is an intentional process that *begins with reasoning* when we seek something or ask a question.

---

8   Bird, *The Divining Hand*, pp. 77–92.
9   Leftwich, *Dowsing: The Ancient Art of Rhabdomancy*, p. 13.
10 Willey, *Modern Dowsing: The Dowser's Handbook;* cited in Lloyd Youngblood, "Dowsing: Ancient History" (http://dowsers.org/dowsing-history).
11 Lutie Larsen, "What Is Intuition?" *The Radionic Homestead Report*, vol. 8, no. 2. March/April. pp. 1–2.

Despite this difference, dowsing and intuition exercise the same basic human faculty, and practicing one seems to enhance the other.

This raises a basic questions of the source for intuitive insights and the answers we get when dowsing. Where do they come from? Most psychotronics practitioners believe that this information, whether retrieved by intuitive insight or by dowsing, comes from a universal energetic field of information. This theoretical field has many names and descriptions, some of which you might know. It has been called the akashic record, the akashic field, the universal mind, the collective unconscious, the enfolded sea of consciousness, and the implicate order.[12]

The Implicate Order

The concept of an *implicate order* has strong scientific origins. It arises from the work of quantum[13] physicist David Bohm (1917–1992). According to Bohm, our universe is much like a large, continuously flowing and ever-changing hologram—somewhat like a 3-D movie, only so much more. Bohm calls this a *holomovement*.[14]

The concepts are hard for anyone to wrap his or her head around, but it helps if we understand a few things about *holograms*. For many of us, holograms are little more than interesting three-dimensional tricks of light we might see in a science museum, or on that expensive belt buckle your cousin bought in New Mexico. To start with, let's be technically correct. The 3-D light image we observe in the museum or wherever is *not* a hologram. The images we see are holographic *projections*. The hologram is actually something like

---

12 Edith Jurka, "Brain Patterns Characteristic of Dowsers: As Measured on the Mind Mirror," *American Dowser*, Feb. 1983, vol. 23, no. 1, pp. 5–11.

13 Quantum theory is the theoretical basis of modern physics that explains the nature and behavior of matter and energy on the atomic and subatomic level. The nature and behavior of matter and energy at that level is sometimes referred to as quantum physics and quantum mechanics (see https://whatis.techtarget.com/definition/quantum-theory).

14 Talbot, *The Holographic Universe*, p. 48.

a photographic plate that holds images in the form of *interference patterns*. When we look at the hologram itself, we see only a chaotic mess—no discernible image at all. However, by shining a laser through the plate, a 3-D projection emerges. The hologram stores the information of the image we see projected.

The fact that holograms hold images and other forms of information as interference patterns is a key factor. But to understand its significance, we need to back up a bit and look at traditional analog still photography. Actually, we should go way back to a time when the negative photographic images were printed on glass. When you looked at one of those glass negatives, you would see the whole image of your subject in reverse. (Later, with additional processing, the negative would be used to print one or more positive images on paper to place on the wall or mount in a photo album.) It would be absolutely correct to say that the glass negative stores information in the form of an *inverse* analog image.

Should you accidentally shatter a glass negative, you have pretty much destroyed the image. You can pick up the shards, but any one of them will reveal only a small portion of the image information the entire plate held.

Unlike analog photographic negatives, holograms reveal nothing to the human eye as we look at them. There is nothing more than patterns of waves that overlap each other. However, once you shine the proper laser light through the hologram, you uncover the image hidden there. Should you shatter the hologram and retrieve only a small shard, you'll be surprised to find that, when the laser is projected through it, the entire 3-D image is still revealed. Information in a hologram, therefore, is held or *enfolded* throughout.

Thus, if Bohm is correct and the universe resembles a holomovement, our material reality (the *explicate*) is a projection of sorts from an underlying reality (the *implicate*). And whereas our material world appears to be a vast collection of separate parts (like the shattered photographic negative), our deeper reality is like

Psychotronics and a Biodynamic Garden

the hologram—"an undivided wholeness of all things; where everything...is made out of the [same] seamless holographic fabric."[15] Essentially, what this means is that there really is no separation of things; everything is interconnected.[16]

This concept is difficult to grasp. We observe separation between ourselves and others, between ourselves and everything else in our world. True, but that is a characteristic of the holomovement projection—the 3-D "reality show" we star in. This reality show is our physical world—the *explicate order*, which is the projection of the implicate order—the underlying reality. Stated a bit more scientifically, we might understand the nature of our daily physical reality as "surface phenomena...forms that have temporarily unfolded out of an underlying implicate order in a dynamic form."[17,18]

Cutting to the chase, I wrote this long, tortuous discussion in an effort to explain that dowsers (and intuitives) are tapping into the implicate or underlying reality of the universe to retrieve information. It makes sense. All information about everything in the physical–explicate world is stored there!

Emanations

While many dowsers believe as I do—that we are tapping into the universal energy field (or implicate order)—not all of them do. Some have another explanation. They believe that they picking up on direct emanations from the objects they seek. This is often argued by those who specialize in finding mineral veins, underground pipes, subterranean water flows, or lost objects. It is easy to understand why.

---

15 Ibid., p. 48.
16 Ibid., pp. 38–48.
17 F. David Peat, "Nonlocality in Nature and Cognition," p. 304; quoted in: Carvallo, *Nature, Cognition and System II*.
18 Talbot, *The Holographic Universe,* pp. 38–48.

If emanations are involved, we might consider them *local phenomena*, and part of the explicate order—the unfolded physical–material world. This is certainly true if they have electromagnetic characteristics that sensitive conventional instruments can also pick up.

While local emanations might play a role when one is dowsing for objects, earth energy lines, and such, they would not provide an explanation for the breadth of dowsing applications, which includes answering posed questions—a circumstance where one is not searching for something that "emanates." In such cases, the logic of accessing a universal energy field holds more sway. So even when local phenomena appear to play a role in certain kinds of dowsing, I expect we are still, ultimately, tapping into the nonlocal reality of the implicate.[19]

Locality and Nonlocality

In the previous few paragraphs, I've tossed around the terms *local* and *nonlocal*. Let me clarify. In quantum physics, *local action* means that an object is directly influenced or moved by something only in its immediate surroundings.[20] A practical example: if you place a billiard ball on a pool table, it will move only if it is pushed, struck by another ball or similarly forced. We call this *causation,* or *cause and effect.* You would have learned about it in high school if you hadn't been goofing off.

Now, suppose that same billiard ball suddenly rolled across the table with no apparent cause. You would be mystified to say the least; you might even spill your beer. But suppose you learned that a player at an adjacent pool table had stolen one of the billiard balls belonging to your set. You then noticed that your ball moved every

---

19   British Physician and noted dowser, Aubrey Westlake, suggested that multiple factors might be at play in dowsing. Still, he argued that an *etheric counterpart* must surely be behind any form of local emanations, and that these would ultimately be the explanation for dowsing phenomena (Westlake, *The Pattern of Health,* p 159).

20   *Principle of locality;* see https://en.wikipedia.org/wiki/Principle_of_locality.

time the player struck the ball he stole. While the odds of this happening are incredibly low, a quantum physicist would be tempted to call it *nonlocal action* (or *spooky action at a distance*).

In nonlocal events, the movements of objects are due to activity occurring elsewhere. The objects might seem unrelated because they are separated in space. However, on the implicate level—in the underlying reality—the two objects are somehow bound to each other.[21] In the example of the billiard balls, the fact that they belonged to the same set explains the relationship. Such relationships exemplify *quantum entanglement*.[22]

While it makes for a good example, we are unlikely to see nonlocal action on the scale of billiard balls. Nonlocal activity occurs chiefly on the quantum level with subatomic particles. This is the level at which we believe that psychotronics functions.

Frankly, we could spend much time here on the topic of nonlocality, but I don't think it would be very useful. Besides, I risk getting in way over my head. But I will note three characteristics of nonlocal action on which physicists agree and that we often observe in psychotronic work. Nonlocal action is, they say:

- unmediated—that is, without anyone or anything intervening or making physical contact;
- unmitigated, in that it is not diminished by distance;
- immediate.[23]

---

21 *Action at a distance;* see https://en.wikipedia.org/wiki/Action_at_a_distance.

22 In quantum physics, entangled particles remain connected so that actions performed on one affect the other, even when separated by great distances. The phenomenon so riled Albert Einstein that he coined the phrase "spooky action at a distance" (Live Science: https://www.livescience.com/28550-how-quantum-entanglement-works-infographic.html).

23 Herbert, *Quantum Reality: Beyond the New Physics*, pp. 212–214.

The Human Antenna and Resonance

So, how do we tap into the universal energy field? How do we retrieve the *energy information* stored there? In the physical world, if we talk about detecting energy information in the atmosphere, we logically think about radio—about antennas, receivers, and related paraphernalia. It turns out that this mental association *is* helpful to our understanding.

Dowsers believe that the human nervous system actually functions much like an antenna and receiver. The average individual has roughly thirty-seven miles of nerves that act a kind of integrated sensor.[24] But the system needs to do more than detect radiations. We need to separate the wheat from the chaff, as it were. Remember, in addition to being immersed in the vast complex universal energy field, we are also bombarded with all manner of local emanations and radiations that include television and radio broadcasts, microwaves, and radioactivity from power stations. We require a mechanism to tune into the specific energy information we want, and eliminate the "noise."

It might seem like an insurmountable task, but thinking about it one realizes we do something akin to this already. Without thinking about it, we focus our five basic senses on specific visuals, smells, sensations, and sounds all the time, while simultaneously shoving everything else into the background. This is what we do when we converse with someone in a noisy café. It's what happens when we focus our eyes on a single bird in a tree that is full of them. It is how we appreciate the taste of shiitake mushrooms on a six-topping pizza. Needless to say, the broad tableau of sounds, images, and tastes don't vanish when we do these things; they are still there. By using our *intent process* or *intentions*, we are able to focus and sort through it all to tune in to something in particular. This is a useful analogy, but we need something more to help us understand how

---

24 Massey, *Alive to the Universe: A Layman's Handbook of Supersensonics*, pp. 23–24.

our *sixth sense* sorts through the universal energy field to find what it wants. For this, we'll return to radio.

If your radio is as old as mine, you still use a tuning dial to locate your favorite stations. You rotate the dial until you find the point of *resonance* with a particular station's broadcast signal. Once the point of resonance is found, the energy information—news, music, drama—flows out of the speaker. The concept of resonance is not difficult to grasp. One of the leaders in our radionics community, Ed Kelly, gives a nice demonstration of resonance using tuning forks. When he strikes a 108-Hz[25] tuning fork, for example, causing it to vibrate, another 108-Hz tuning fork, situated nearby, spontaneously vibrates in *sympathetic resonance*, even though the two are not touching. Other tuning forks of different hertz, do not vibrate because they are not in resonance. When dowsers focus on something—a question, an object, another person—they establish resonance with it and set up circumstances where energy information can flow.

In *Alive to the Universe*, Robert Massy explains it this way:

> We know that every particle is sending its radiations throughout space, crisscrossing through the spaces in our bodies. All the information of the universe is with us at once. How do we then pick out the part that interests us? We do so by remembering that we are also sending out waves from our bodies and minds, which are constantly interacting with this universal wave-field. When we tune our own wave-field to the wave-field of a particle, an enhanced effect is set up called resonance. It is this resonant signal that we detect.[26]

The resonances we seek in dowsing are much more complex than radio waves and tuning forks. Imagine the complexity of an individual corn plant; its frequency is made up of the vibrations of all the atoms, molecules, and whatnots that comprise it. So, how we might achieve resonance with a corn plant, a field of corn plants,

---

25 *Hertz (Hz)* is a unit of frequency equal to one cycle per second.
26 Massy, *Alive to the Universe*, p. 27.

an animal, or a person is difficult to fathom. Yet, our nervous system seems to have that capability. By employing psychotronics we can enhance that capability to tease out the specific resonances that make up the whole corn plant; we can assess its nutrient content, moisture status, amino acid balance, and pest vulnerability. This sort of thing is routine in radionics, but we can also do it by dowsing.

Intuition and Intuitives

Earlier I wrote about intuition and its relationship to dowsing. It is worth saying a bit more. There *are* people whose intuition is especially strong. We sometimes refer to such folks as psychics or intuitives. Intuitives readily tap into the universal energy field with little apparent effort; they "just do it."

I suspect most of us have had one or more experiences in which information or images from the universal energy field leaked through to our waking consciousness. We call these paranormal experiences, psi episodes, or *extrasensory perception* (ESP). The average person seldom knows what to think when confronted with such events.

Ingo Swann (1933–2013), noted psychic, artist, and researcher, had this to say about human impediments to ESP and psi experiences and the failure to understand them in useful ways: "The greatest drawback to any progress in comprehending extrasensory perception at the individual level and in parapsychology is trying to make ESP fit into the reality we think is the only reality."[27]

Most of us are not natural intuitives. I'm certainly not! But intuitive skills can be learned and enhanced. Dr. Aubrey T. Westlake (1893–1985), the prominent British authority on radiesthesia and alternative medical therapies, refers us to Rudolf Steiner's *How to Know Higher Worlds* as a practical guide for those with the perseverance and devotion to seek direct experience of higher realities. While encouraging, he describes this as a long hard road, and that,

---

[27] Swann, *Everybody's Guide to Natural ESP: Unlocking the Extrasensory Power of Your Mind*, p. 46.

practically, most us should pursue physically mediated means such as dowsing to tap into the universal energy field.[28]

We can conclude, then, that dowsing is a means to use and manage our ESP potential. We use our *intent process*, through either seeking or asking questions, to sort through the vast amount of energy information that permeates the universe.[29] But, unlike natural or trained intuitives—who can "see" or otherwise draw meaning from direct extrasensory experiences—most dowsers must rely on our body's physiological cues, which we call *dowsing response(s)*.

The Dowsing Response

When L-rods cross, a forked twig dips, or a pendulum changes its swing, dowsers know that either they have found what they sought—a mineral vein, an underground stream, a lost object, and so on—or they have the correct answer to a question. These visible actions, many and varied, are commonly called *dowsing responses*.

Most dowsers believe dowsing responses are *ideomotor reflexes*—unconscious, involuntary physical responses to ideas or other stimuli.[30] A somewhat extreme but obvious example of an ideomotor reflex is our body's response to physical pain. We don't think about it when we yank our finger away from a candle flame or a hot stove. Our body just reacts, as it should.

The ideomotor reflexes dowsers experience are much more subtle. They manifest as slight involuntary muscular or nervous system reactions, so slight that we are unaware of them. This is where simple tools like L-rods and pendulums play their role; they amplify these tiny neuromuscular movements so that we more readily recognize them.

---

28 Westlake, *The Pattern of Health: A Search for a Greater Understanding of the Life Force in Health and Disease*, p 157.

29 I don't mean to imply that practicing dowsing eliminates other kinds of psi experiences. Quite the contrary. The more one dowses, the more one is likely to experience other forms of psi phenomena.

30 See en.wikipedia.org/wiki/Ideomotor_phenomenon.

Psychotronics

One of the earliest attempts to study the dowsing response was conducted by a seventeenth-century Jesuit priest, Athanasius Kircher (1602–1680). His findings supported the ideomotor reflex hypothesis.[31] Further research in the nineteenth century by the French chemist Michel Eugène Chevreul (1786–1889) concurred with Kircher.[32] In more recent times, we have similar support from studies of *galvanic skin response*.

Galvanic skin response (GSR) is another form of ideomotor response. GSR is a change in the electrical resistance of the skin; it is one of the parameters measured by polygraph and biofeedback instruments. The variation in electrical resistance results from involuntary reactions in our sweat glands and changes in skin pore size caused by altered emotional states.[33] Researchers sometimes refer to this as "emotional arousal."[34] Like other forms of dowsing response, GSR is a very subtle reaction—one that most of us would not, generally, be conscious of. That's why sensitive GSR instruments are employed. They serve as detectors and amplifiers. While I'm not aware of any dowsers who use GSR, it has been used in radionics—something I'll discuss later.

The ideomotor phenomenon is the current and most scientific rationale I know of to explain the dowsing response. There are some practitioners, however, who will insist on psychokinetic[35] or other more mysterious explanations. Their claims are *not* without foundation.

---

31 Steven G. Herbert, "A brief history of Investigations into the Mechanisms of Dowsing," *American Dowser,* spring/summer 2011, vol. 51, no. 2–3, p. 55.

32 Ibid., p. 57.

33 Anon., *The Psi Connection: A Builder's Handbook of Psychic Equipment,* The Heritage Institute, Plainfield, Wisconsin. p. 29.

34 See en.wikipedia.org/wiki/Electrodermal_activity.

35 Edith Jurka, "The Mind Mirror and the Dowsing Response," *American Dowser,* spring 2005, vol. 45, no. 2, p. 37. *Psychokinesis* is the movement of physical objects using the mind and without physical means.

Baron von Reichenbach (1788–1869), the noted chemist, naturalist, and philosopher who is identified with the concept of *Odic force*,[36] experimented with pendulum movements. He devised an apparatus that shielded the pendulum from involuntary neuromuscular (ideomotor) reactions, while still allowing physical finger contact with the string suspending the *bob*.[37] His trials confirmed the hypothesis that ordinary people are *unable* to move the pendulum by psychokinetic means, thus reinforcing the notion of ideomotor action. However, when he tested known "sensitives," they were capable of causing the pendulum in his mechanism to react.[38] Von Reichenbach's findings were duplicated by another contemporary, F. de Brichec, using a similar apparatus.[39] So there *is* substance to the claim that the dowsing response might, in the case of some gifted individuals, be psychokinetic. For most of us, though, the ideomotor reflex is the likely cause and explanation for our dowsing responses. With that said, it is increasingly evident that the vast majority of us can dowse, if we're willing to learn. Again, as the author Robert Leftwich put it: *it is our birthright.*[40]

Response Training

Sometimes, the meaning of a dowsing response is quite clear. If you are dowsing a well site and your willow branch dips downward, you might assume you've found the correct location. It is less clear when you are holding a pendulum and asking a yes or no question.

---

36 *Odic force* describes a hypothetical vital energy or life force identified by Baron Carl von Reichenbach, who coined the term in 1845 from the Norse god Odin; see en.wikipedia.org/wiki/Odic_force.

37 The term *bob* is defined in *Merriam-Webster* as a hanging ball or weight.

38 Tansley, *Dimensions of Radionics: Techniques of Instrumented Distant-Healing*, p. 181.

39 Steven G. Herbert, "A Brief History of Investigations into the Mechanisms of Dowsing," *American Dowser*, spring/summer 2011, vol. 51, nos. 2–3, p. 57.

40 Leftwich, *Dowsing: The Ancient Art of Rhabdomancy*, p. 13.

Pendulums can swing back and forth, side to side, and rotate in opposite directions. Which movement indicates "yes," and which one indicates "no?"

When it comes to the pendulum, I have a simple but reliable self-training method that I use when teaching novices. I start by placing a C- or D-sized battery on a flat surface with the positive end up and suspend my pendulum over it. If the pendulum does not begin to rotate in a clockwise direction, I nudge it in that direction. I then flip the battery over and hold the pendulum over the negative pole; it should now begin to rotate counterclockwise. From then on, I interpret the positive or clockwise pendulum rotation as a "yes" response, and the negative, counterclockwise rotation as my "no" response.

When I First Learned to Dowse

I actually learned L-rod dowsing several years before I studied radionics. I was in rural New York State in the mid-1980s attending an organic farming workshop. One of the coordinators, David Yarrow, an experienced dowser and author of the booklet *Dowsing History and Techniques*,[41] took the time to teach me. This was an exciting and memorable experience. Despite that, I wasn't motivated to pursue dowsing further. And, strangely, when I began studying radionics, I really didn't grasp how the two activities were related.

A Dowsing State of Mind

It should be no surprise that practitioners enter a sort of mental state or "zone" when they dowse. It is a focused activity and one's intentions must be clear. We are fortunate to have some insights into this mental state thanks to the work of biophysicist and psychobiologist Dr. C. Maxwell Cade (1918–1985) and his collaborator Geoffrey Blundell (1923–2003), an electronics designer. The two developed a special electroencephalograph that

---

41 Yarrow, *Dowsing History and Techniques*, p. 22.

objectively observes changes in mental states during meditation and other altered states of consciousness. Called the *mind mirror*, the instrument records brain frequencies of the left and right brain hemispheres simultaneously.

Using the mind mirror, Cade and Blundell observed that human brains emitted mostly theta waves in the low, four- to seven-hertz range during dreaming sleep, and even lower frequency delta waves (0–4 Hz) when in deep, dreamless sleep. By contrast they emit fast alpha waves (8–13 Hz) during a relaxed but awake state, and even faster Beta waves (13–30 Hz) when fully awake and alert. The researchers found that when one used transcendental meditation™ techniques the brain emitted waves predominantly in the alpha and theta ranges—as one would when relaxed and/or dreaming. Zen meditation practitioners produced similar alpha and theta patterns, but also emitted beta waves reflecting the "here and now" awareness associated with Zen practice.[42]

In 1982, Dr. Edith Jurka (1915–2012) brought a mind mirror to an American Society of Dowsers (ASD) conference to measure the brain wave patterns of seven gifted dowsers in attendance. To her surprise, the mind mirror patterns most closely resembled those generated by Zen meditation, but also showed significant output of the delta waves produced during deep sleep! Accompanying this were higher emissions of beta waves as well (see graphic).[43] According to Jurka, Cade believed that the delta frequencies correlate with reaching out to the unknown, whereas the enhanced beta waves reflect simultaneous attention to the outside world.[44] Truly a remarkable state of mind! As longtime dowser, Ed Stillman wrote in 1998:

---

42 Edith Jurka, "Brain Patterns Characteristic of Dowsers: As Measured on the Mind Mirror," *American Dowser*, Feb. 1983, vol 23, no. 1, pp. 5–11.

43 The graphic is originally adapted from Edith M. Jurka, "Brain Patterns Characteristic of Dowsers," *American Dowser*, fall 1991, pp. 8–14.

44 Jurka, "Brain Patterns Characteristic of Dowsers: As Measured on the Mind Mirror," *American Dowser*, Feb. 1983, vol. 23, no. 1, pp. 5–11.

With the dowser's "global" expansion of high-power brainwaves simultaneously in delta, theta, alpha and beta, dowsing is truly a creative state of mind, a creative altered state which includes actively asking questions and receiving answers through physiological body pathways.[45]

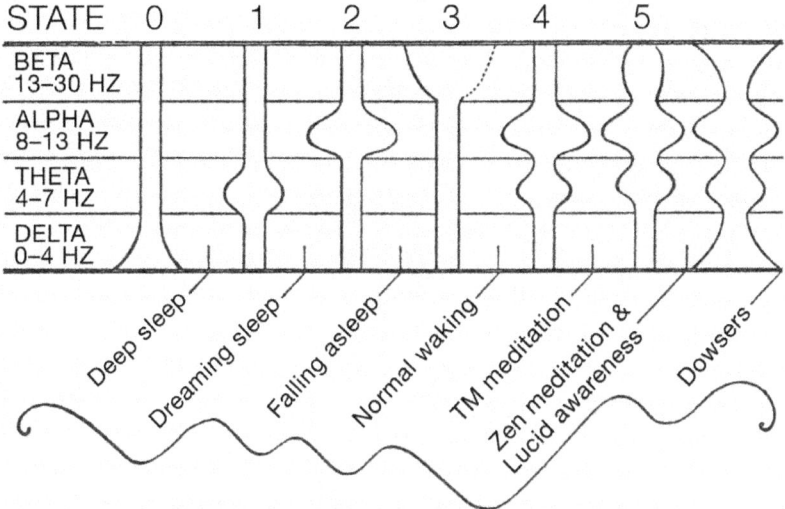

Figure adapted from articles in Dowser's Journal

Moving on to Radionics

Unlike dowsing, radionics immediately and thoroughly captured my imagination, even though it was harder to learn. The instrumentation had a lot to do with it. Instruments make radionics feel much more scientific and palatable to those of us still struggling with the conventional scientific paradigm. There is also a universal truth that "boys love toys."

---

45 Ed Stillman, "Dowser's Brainwave Characteristics, Part Two: Brainwave Coherence and Delta Waves," *American Dowser,* spring 1998 vol. 38, no. 2, p. 24.

Defining Radionics

Radionics has many definitions; most tend to be off the mark or just wrong. Several of the really bad ones come from detractors. Sadly, this nonsense comes with the territory.

Fortunately, there are some fine definitions as well. One of the better ones comes from author and practitioner David V. Tansley, DC (1934–1988): "Radionics is a system of distant diagnosis and treatment, which utilizes the human faculties of extrasensory perception in conjunction with certain specially designed instruments, to determine the underlying root causes of disease in a living organism."[46] This definition highlights the medical origins of radionics,[47] as well as Tansley's credentials and interests. It is concise and gives one a good sense of radionics. It highlights the ESP connection and thereby the link to psychotronics and dowsing as well.

Because I'm an agronomist and horticulturist and not a medical practitioner, there is another definition I particularly like that is used by a friend and fellow practitioner Ron Barone. It hints at the wider range of practical radionics applications: "Radionics is the practice of focusing intent.... In order to determine almost anything and to be able to broadcast continually or intermittently that intent across time and space to anyone or anything anywhere, past, present, or future, without having to maintain focused thought intent."[48]

Barone's definition tells us that the first objective of radionics is identical to that for dowsing—finding information. In fact, while tools and procedures differ, radionics and dowsing are closely allied. Virtually everything you have read so far in this chapter

---

46 Tansley, *Radionics: A Patient's Guide to Instrumented Distant Diagnosis and Healing*, pp. 1–2.

47 Radionics, as we recognize it today, was first discovered by a physician, Dr. Albert Abrams, who used it for the diagnosis and treatment of human diseases (Tansley, *Radionics*, pp. 7–22).

48 Ron Barone, "Radionics: Taught by the Master Teachers Symposium," Rapid City, SD, Oct. 2016, pp. 27–29,

about dowsing applies to radionics as well. This is especially true as regards discussions on the universal energy field, the holographic nature of reality, resonance, intuition, and dowsing responses. Furthermore, radionics generously weaves dowsing procedures into its routine operations and procedures. Some have even taken to calling radionics "instrumented dowsing."

Radionic Instrumentation

Complex instruments are the most visible characteristic of radionics; they are what distinguishes radionics most from dowsing. People find the instruments intriguing, and they are often what draws many scientists and engineers to radionics. Many radionics instruments resemble radios and other electronic devices from the early to mid-twentieth century. Still others are integrated with computers and have a more modern look. Most feature recessed wells, antennas, dials, wires, and switches. They have a *reaction plate* which enhances the dowsing response and doubles as an antenna for radionics *broadcasting*. Broadcasting is the preferred term for *treatment at a distance*—another distinguishing characteristic of radionics.

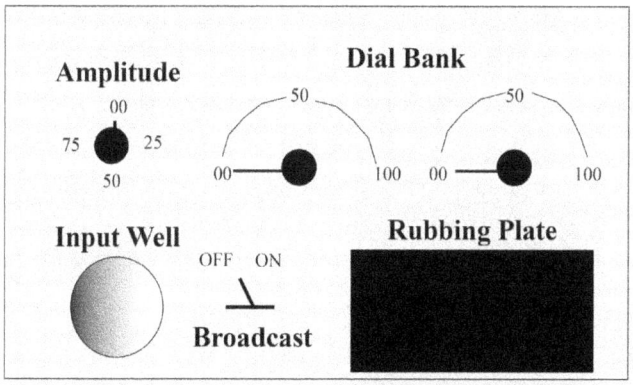

Basic Radionic Instrument
Based on the Hieronymus System

I see the role of radionics instrumentation as threefold:
1. Instruments enhance the dowsing response.
2. They help clarify and hold the intent of the operator throughout the analytical or "questioning" process.
3. They also retain and amplify the operator's intent for sustained broadcasting.

Radionic Instruments and the Dowsing Response

Like dowsing, radionics relies on dowsing responses to tell us when our findings are correct. I know of only three response methods: pendulums, galvanic skin response, and finger-sticking.

There a several radionics systems or methods. In several of them, operators use a pendulum much the way dowsers do, but in conjunction with an instrument. The British radionics pioneer Malcolm Rae (1913–1979) preferred to use a pendulum. He wrote that the pendulum had an advantage over the finger-stick reaction *"because it is less liable to be influenced by changes in the practitioner's skin moisture, by fatigue, or by the humidity in which the practitioner is working."*[49]

At least one radionics instrument design employs galvanic skin response (GSR). It is described in a slim book entitled *The Psi Connection*.[50] *The Psi Connection* was likely written in the early 1980s, possibly by Dr. W. E. Davis, the author of another widely distributed book on radionics titled *The Black Box*.[51] The use of GSR is also described in a video from the same Wisconsin-based publishers, titled *Radionics: Science of Tomorrow*.[52] I know of no one currently making such instruments or actively promoting the idea.

---

49 Malcolm Rae, "Radionic Instruments and Rates," *The Radionic Quarterley,* in Tansley, *Dimensions of Radionics,* p. 73.

50 Anon., *The Psi Connection: A Builder's Handbook of Psychic Equipment,* The Heritage Institute, Plainfield, WI, p. 29.

51 W. E. Davis, *The Black Box and Other Psychic Generators,* The Heritage Institute, no date, Plainfield, WI, p. 68.

52 *Radionics: Science of Tomorrow* (VHS tape). The Heritage Institute, Plainfield, WI.

It seems that most practitioners prefer the *finger-stick* for amplifying the dowsing response. The *finger-stick* is a subtle sensation that the operator experiences while stroking the smooth surface of a reaction (or rubbing) plate. The sensation arises from behavioral changes in the sweat glands of one's finger pads[53] and seems akin to galvanic skin response.

A further explanation for the stick plate phenomenon is suggested in a 1953 Science article by Edward Mallinckrodt and associates. The researchers reported a dragging sensation on their fingertips when stroking "a smooth metal surface covered with a thin insulating layer...[while] the metal is connected to the ungrounded side of a 110-volt power line. They describe these as electrically induced vibrations by highly sensitive cutaneous receptors in the skin."[54]

While the Mallinckrodt findings are valid and add to our understanding, they do not supplant any previous theories or hypotheses. First of all, Mallinckrodt's observations were made only when electricity was flowing through the experimental apparatus. Not all radionics instruments with stick plates use electricity, yet operators are able to get a stick. Likewise, many of us dowse merely by stroking any available smooth surface and get a stick. And it seems to make no difference (to me, at least) whether those surfaces are made of wood, metal, stone, or plastic.

Witnesses

Witnesses are specimens obtained from the subject(s) we propose to analyze or target with a broadcast. Traditional witnesses include saliva, blood, hair, or fur. When working with plants, I often use leaves, seeds, twigs, bark and soil. Witnesses are placed

---

53 Rae, "Radionic Instruments and Rates," *The Radionic Quarterly*; reproduced in Tansley, *Dimensions of Radionics*, p. 73.

54 Edward Mallinckrodt, A. L. Hughes, and William Sleator Jr, "Perception by the Skin of Electrically Induced Vibrations," *Science*, vol. 118, Sept. 4, 1953, pp. 277–278.

within recessed wells or specialized plates that are either built into the instrument or connected to it by wires.

Witnesses enhance our focus and prolong our connection with the subject for as long as needed. Witnesses work hand in hand with radionics instruments, which are designed to hold and sustain our intentions.

The rationale for witnesses derives from the concept of *quantum entanglement*, as discussed earlier in this chapter. A subject's witness is *entangled* with the subject on the implicate level, therefore they *resonate*. This provides the energetic link that radionics practitioners need to establish.

An important thing to understand about witnesses is that they provide ongoing, real-time energetic links to their subjects. For example, I can use a leaf witness I gathered from a tree six months ago to determine its vitality today. This is what distinguishes witnesses from *samples*, such as those we use for chemical analysis. A soil or plant sample captures the chemical and physical state of the subject only at the time it is collected.

Rates

Rates are another common characteristic of most radionics systems. Rates are usually number codes that we set on the instrument dials to specify what we are investigating. If placing a witness in the instrument isolates and defines the subject we are looking at, setting a rate narrows our focus to those things about the subject that we want to study and measure. When working with soils and plants, I use rates for nutrients, plant functions, soil organisms, diseases, pollutants, and much more.

We also have *reagent* rates. Reagents are usually substances including color filters, gemstones, minerals, vitamins, and fertilizers that we might use in an analysis or to support radionics broadcasts. It is customary for many practitioners to keep a wide variety of reagents in lead-free glass tubes, organized and readily available. I used reagents like these in the 1980s when formulating foliar fertilizers for blueberries and other crops. I kept, perhaps, fifty or more tubes of fertilizers, plant vitamins, amendments, and surfactants at hand.

I still use physical reagents, but these days, I am more inclined to use reagent rates instead. For example, suppose that my tomato plant needs a boost. I suspect that seaweed extract might do the trick, but don't have any on hand. I can, first, radionically measure my plant's vitality. Then, by doing further analysis with the common rate for seaweed extract (52.50–77.00),[55] I can determine whether I was correct or not. If I was correct, I can use the same rate to broadcast the essence of seaweed extract to my plant.

A few radionics systems, such as Malcolm Rae's, use specially made cards in place of rates. Like rates, these cards define the specific things the operator wants to measure, evaluate, and bring into balance. There are also cards that substitute for reagents, in the same manner as radionics rates.

An important point about rates: the numbers, themselves, are not frequencies. They are codes—codes that aid us in accessing complex patterns of frequencies for organs, systems, functions, characteristics, molecules, or whatever it is that we are investigating. I point this out to alleviate the dismay novices feel when they discover that fire blight, for example, has two or more analysis rates, or that *any* single rate might refer to more than one thing. This might seem contradictory until we remind ourselves that rates, witnesses, and the instruments themselves are tools to clarify and support our intent process. Therefore, we can and should bend them to our needs.

---

55 Rates for common fertilizer reagents are available from several sources. A substantial list is provided in a later chapter of this book.

A Brief Example of a Radionics Analysis and Broadcast

What follows are the procedures I would typically use for the analysis of a plant, followed by a broadcast. For this example, I'm choosing elderberry, a promising new crop in our region.

I begin by obtaining a suitable witness. Using a fresh pair of nitrile gloves and a clean scissors, I collect a leaf from one of my elderberry plants and place it in a glassine envelope.

Before inserting the witness into the instrument, I neutralize the input well to remove any stray energy information that might remain from previous tasks. Many Hieronymus-style instruments have a toggle switch for this purpose. Mine does not, so I use a small horseshoe magnet. I insert it into the well and sweep across and around the outside casing of the instrument.

After placing the witness in the input well, I dowse to confirm its integrity, asking: "Is this witness suitable for analyzing and broadcasting to the elderberry plant?" I dowse this question while either stroking the instrument's rubbing plate or observing the reaction of a pendulum that I suspend over it.

I'm doing an abbreviated analysis that focuses specifically on plant *energy centers* or *chakras*,[56] and the dominant plant structures associated with each. These concepts and the form that follows are based on the work of one of my mentors, Lutie Larsen.[57]

Forms, commonly called *analysis sheets* are frequently used in radionics. Like this one, analysis sheets usually feature a number of preselected rates that are organized in some common-sense manner. The first rate, *Plant Vitality*,[58] is a baseline measurement—a sort of "average value" we use for comparison purposes. It is succeeded by the four plant chakra rates, each of which is followed by one or more

---

56   *Chakras* are the focal points on the etheric body for the reception and transmission of energies (Tansley, "Radionics and the Subtle Anatomy of Man," C. W. Daniel, Essex, UK. p. 23).

57   Lutie Larsen, "Tips and Techniques for Farmers: Wise Woman Ventures" (993 West, 1800 North, Pleasant Grove, Utah, 2008), pp. 132–136.

58   The same rate (09-49) is used broadly in radionics analyses as a baseline. It is commonly referred to as *General Vitality*.

Psychotronics

aspects of the plant that are closely associated with or "governed" by that energy center.

Hypothetical Analysis of Elderberry Plant

Rate	Description	Measurement	Notes
9–49	plant vitality	523	
48–50	survival center*	530	
3–32.5	terminal bud	535	
25–57	energy support center*	498	
7.5–14	phloem	494	
5.75–3.25	stomates	489	
30–56	physical support center*	280	low, out of balance
4.75–3	xylem	250	low, out of balance
14–44	root center*	540	
38–22.5	root	530	

*plant chakra

The Figure depicts analysis of Plant Vitality:
- the leaf witness (W) is in the input well
- the rate 9-49 is set on the dial bank
- the operator finger-strokes the rubbing plate
- the amplitude dial is rotated until a stick is felt (at 523 or 52.3)

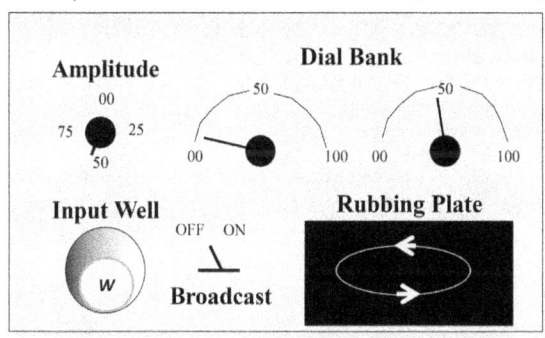

I enter each rate separately on the rate dials, then rotate the intensity dial while gently stroking the rubbing plate. Once I feel a drag or *stick* on the rubbing plate, I stop and record the measurement indicated by the intensity dial. I then proceed to set the rate dials to the next rate and get another reading. The preceding graphic details the process for getting the measurement for plant vitality.

Psychotronics and a Biodynamic Garden

Once I've completed the entire analysis, I study the results. Using the Plant Vitality reading as a baseline, I look for any readings that vary from it by ten percent or more. In this instance, the physical support center and the xylem are both out of balance. I would then check rates for diseases, water stress, environmental pollution, and similar factors to find the cause(s) behind the imbalances.

Taking the example to the next stage, let's assume I entered the common rate for Verticillium wilt (13.5–18.6), and my reading indicates the plant had a serious infection. I would certainly check the horticulture literature for recommendations, but I would also expect to use radionics broadcasting.

Typically, the broadcast would entail leaving the plant witness in the input well and resetting the rate for Verticillium on the rate dials. I would dowse to determine whether there were any supporting reagents I might include. It would be no surprise to find that elemental sulfur is indicated.

I would place a reagent tube of sulfur alongside the witness in the well, dowse to confirm "it is safe and advisable to broadcast," and then toggle the broadcast switch "on" (see graphic). Finally, I would dowse to determine how long the broadcast should be maintained.

A Broadcast to Suppress Verticillium Wilt:
- the leaf witness (*W*) is in the input well
- the rate for the fungus 13.5-18.6 is set on the dial bank
- the operator dowses for a supporting reagent (sulfur)
- after confirming safety & advisability, broadcast is toggled on
- dowse for time

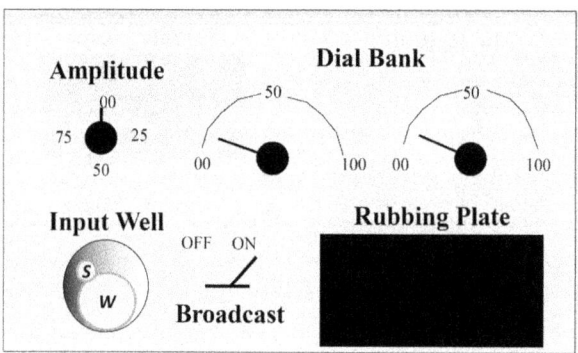

Subtle Anatomy and the Seven Bodies

Bohm's holomovement theory helps us understand the nature of reality. It provides an explanation for the universal energy field (the implicate order) and its relationship to the tangible physical world (the explicate order). Still, it leaves a huge gap in our understanding of reality. To wit: *how* does the implicate order craft the explicate order? How is the physical aspect of the universe created from the energy information held in the universal energy field? We know that holographic projections are created when we shine the correct laser beam through the hologram. But this analogy falls far short of answering question of how we and our physical world got here. Like it or not, this is where cosmology enters the picture.

Cosmology deals with accounts and theories on the origin of the universe. Genesis contains an accounting of creation that is the cosmological touchstone for the Judeo-Christian world. Interpretations of this story vary. Whereas some understand it as emphasizing separation of the physical world from the spiritual, others view the physical world as grounded and linked to the spiritual. My working paradigm borrows from the latter view. I find it consistent with my spiritual beliefs, my experiences with psychotronics, and what I personally see as sensible in light of holomovement theory.

Bastiaan Baan writes,

> ...our foundations are indeed in the spirit...we are rooted in the spiritual world. In the early Middle Ages, it was sometimes depicted literally; in some medieval cathedrals there are images of a human being, turned upside down, standing with his feet in the clouds and with his head pointing to the earth. That portrayed the origin of the human in the spirit, that we have our foundations in the spiritual world."[59]

There is a widely accepted model that describes the descent of the physical from the spiritual in terms of seven vibratory *planes*.

---

59 Baan, *Lord of the Elements: Interweaving Christianity and Nature*. p. 72.

We often refer to these planes as *bodies* or *subtle bodies*, and reference them mostly when speaking about the *subtle anatomy* of human beings.

The seven-planes model is hierarchical. It begins with the highest vibratory plane, the *Divine*, and descends through lower vibratory bodies called the *monadic, atmic, intuitional, mental, emotional,* and *physical–etheric*.[60] At the risk of oversimplifying, the seven planes model describes the physical world as "precipitating" from the higher, finer vibratory states. During this process, the frequencies are stepped down, in much the same way as the voltages in long distance transmission lines (almost 800,000 volts) are reduced to 120 volts upon reaching our homes. The three lower-vibrational bodies (mental, emotional, and physical–etheric) are referred to collectively as the *personality bodies*.[61]

What is the Aura?

To clarify one point: the bodies are not the same as the *aura* or *health aura* which is readily visible to intuitives. Auras result from outward radiation whereas the bodies are "pre-physical" levels.[62]

While it is often discussed as if it were a single plane,[63] the physical–etheric body has two aspects—the *dense physical body*, which

---

60 Terms vary. For example, the *Monadic* is sometimes called the "soul body; *atmic,* the will/spirit; the *intuitional,* the causal; the *emotional,* the astral.

61 Paulson, *Kundalini and the Chakras: Evolution in this Lifetime,* pp. 35–46. Please note, radionic and dowsing work is done almost entirely on the levels of the personality bodies (Tansley, *Dimensions of Radionics,* p. 28). Rudolf Steiner (founder of Anthroposophy and the inspiration behind biodynamics) also described seven bodies, or levels—physical body, etheric (life) body, astral body, "I," spirit self, life spirit, and spirit body (see "The Makeup of the Human Being," in Steiner, *An Outline of Esoteric Science.*

62 Tansley, *The Raiment of Light: A Study of the Human Aura,* p. 44.

63 With specific relation to human subtle anatomy, Aubrey Westlake wrote: "[At] the present time in the evolution of mankind, the etheric body has sunk so deeply into the physical body and is so firmly attached that for all practical purposes they are identical, and this state of affairs maintains

is tangible, and the *etheric* or *vital bioplasmic body*.[64] The etheric body, sometimes called the *etheric double*, is a dynamic, archetypal[65] template on which the dense physical body forms. The closest analogy to this is *electroplating*. In electroplating, electricity causes ions of one metal, such as gold, to attach itself to another metal and conform to its shape. Similarly, in nature, physical matter is drawn to and shaped by the etheric field.

What We Need to Know about the Subtle Bodies and Psychotronics

Rudolf Steiner wrote:

> The whole of esoteric science must spring from two thoughts that can take root in each human being...these two thoughts express facts that can be experienced if we apply the right means.... The first of these thoughts is that behind the visible world there is an invisible one, a world that is temporarily concealed, at least as far as our senses and sense-bound thinking are concerned. The second is that by developing human capacities that lie dormant in us, it is possible to enter this hidden world.[66]

This quotation helps us get to the heart of the matter. For the purposes of psychotronic work, we need to know:

1. There are subtle levels of reality behind our physical world. Our physical world derives from these realms; they constitute a *pre-physical* reality.

---

under all ordinary conditions of life and health" (Westlake, *The Pattern of Health*, p 160).

64 Tansley, *The Raiment of Light: A Study of the Human Aura*, p. 44.
65 An archetype is the original pattern or model from which all things of the same kind are copied or on which they are based; prototype (*Merriam-Webster's Collegiate Dictionary*).
66 Steiner, *An Outline of Esoteric Science*, p. 19.

2. We have various means for studying and working within these pre-physical planes. The one we address in this book is psychotronics.
3. The subtle planes most pertinent to our work with plants are the physical–etheric and the astral.

The Limits of Theories, Analogies, and Paradigms

Over the years I've read numerous popular articles claiming that quantum theory proves that radionics works. This and similar assertions are just not true. While quantum theory provides explanations for how radionics and dowsing might work, it does *not* provide "proof"!

In writing this, I might appear to be undermining my working paradigm. Absolutely not! It is, after all, a "working paradigm," supported by my best understanding of numerous scientific theories and metaphysical concepts.

I acknowledge—as you should—that:

- Theories are just theories! They are ideas about how things are and how they work; they try to explain reality; they might make good sense, but many are not fully proved. In his 1985 book, *Quantum Reality*, Nick Herbert discusses eight theories that attempt to explain reality as described by modern quantum physics.[67] We might choose one or two that support our own concepts, but as time and research advance, that list will change. It might grow, or it might shrink. Our preferred theories can easily find their way into the dustbin.

- Analogies are just analogies! An analogy is used to illustrate a "partial" similarity between two things, the place or places where they correspond to each other. When we say that the heart is like a mechanical pump, for example, we are making an analogy. In relating the hidden implicate order to the explicate, Bohm uses the analogy of holography to explain his theory. His analogy, like others, goes only so far. If we stretch an analogy too far and try using it to explain every aspect of a theory, we risk tearing its fabric to shreds.

---

67 Herbert, *Quantum Reality: Beyond the New Physics*, pp. 15–29.

- Paradigms are just paradigms! Remember that these theories and models constitute *my* working paradigm. While it makes great sense to me, it might not work very well for you. It might not suit you at all!

Honesty and Self-deception

Many dowsers and radionics practitioners assert that there is no limit to the information that we can access from the universal energy field. We just need to ask the right questions. In theory, I accept this. On a practical level, however, I don't think it's quite that simple. Our personal and collective paradigms, our levels of spiritual development, and our personal interests and prejudices can all block access and even produce misinformation. Furthermore, there are limitations on the energy information that each of us is "allowed" to access. These barriers arise not so much because we lack intellectual capacity, but because we lack spiritual development. This is expressed by the theosophical concept of *ring-pass-not*:

> On a spiritual level, if you believe that you are separate from others and alternate levels of existence, a ring-pass-not contains your consciousness and prevents other things from getting in. As you spiritually evolve, and the seeming separations fade away, the ring-pass-not extends, allowing the mind to accept more and more. This is sometimes called the "unfolding of consciousness."[68]

It is not always easy in psychotronic work to recognize whether we are discovering new truths or simply lying to ourselves. There is great risk when we use dowsing or radionics to satisfy and support preconceived notions, prejudices, and self-serving desires. To avoid such traps, we need to be brutally honest with ourselves and willing to self-evaluate. Over time, I've learned there are things I should

---

68 See https://www.llewellyn.com/encyclopedia/term/Ring-Pass-Not.

not know and ought not to ask about. When I'm ambivalent, conflicted, or too emotional about a question or topic, I do not dowse or analyze it radionically. The chances of generating or propagating misinformation are too great.

The Role of Conventional Knowledge

What is the role of conventional knowledge in radionics and dowsing? I've met a few practitioners who eschew conventional study and information. They argue that it will prejudice or compromise their results. I hesitate to judge this approach, because it may work well for these individuals. However, I approach things differently.

I'm inclined toward the view of Dr. Aubrey Westlake, who insists that we *must* have actual information on the subjects we investigate. Westlake believed that without such knowledge we can't know what questions to ask or the best way to ask them. She writes:

> This [applying study of applicable disciplines to radionics] is the correct use of intellectual knowledge; it should provide a reservoir of facts and information from which the intellect can gather material to formulate the right question. But beyond this one must have the ability to reflect, to give proper time, thought and consideration to thinking out in detail any given problem and formulating it precisely. The possession of the faculty is no excuse for poverty of facts or intellectual laziness.[69]

When applying psychotronics to farming and gardening, knowledge of agricultural theories and methods is certainly valuable, as is the study of ecology, nutrition, and related topics. For example, it requires only a rudimentary understanding of crop nutrition to know that plants require phosphorus. It is reasonable, then, to dowse and find that a crop is deficient. It requires a deeper knowledge, however, to know that phosphate is available

---

69 Westlake, *The Pattern of Health*, p 167.

in many forms, each with a different chemistry and soil reaction. Thus informed, a dowser or radionics practitioner would take the additional step of determining which form of phosphorus is best-suited to the soil and crop in question. If you ask a chemical salesperson, he or she will unquestionably tell you to buy whatever they happen to sell.

Permission

A discussion of honesty and self-deception naturally leads to the topics of personal stewardship and permission. I'm referring to practical, moral, and ethical limitations on whom or what you may investigate and influence using psychotronics. It is about boundaries and when you have permission to work beyond them.

Responsible psychotronics practitioners recognize boundaries. They understand that unless matters lie within your stewardship (yourself, your family, your animals, land you own, etc.) you must ensure that you not barge uninvited into another's physical and spiritual space.

It is customary for dowsers and radionics practitioners to ask three questions before doing subtle energy work on someone else's patch.

- *Can I proceed?* This question is used to assess your ability to investigate accurately without risk to yourself or another. It tells you when you are getting in over your head.
- *May I proceed?* Asks whether you have implicit permission to what you plan to do. Even when your intentions are well intended, don't assume you have carte blanche to mess with another on any level. This also applies when a third party, such as a spouse, sibling, or neighbor, makes a request on another's behalf.
- *Should I proceed?* Although you might have the capability and permission to do subtle energy work, there is always a larger

picture to which you are not privy, and you might find that it should go no further. This happens to me occasionally. I don't bother to question why I'm being turned away; I just assume it is in my or the subject's best interests.

2.

How to Dowse

This chapter begins with a quick tutorial for dowsing questions that require yes or no answers. We can adapt almost any dowsing tool for this, including L-rods and bobbers. However, we'll use the pendulum and a further technique called *finger-stick dowsing*.

Choosing a Pendulum

Dr. Linda Lancaster advises using a stainless-steel pendulum, because it does not readily absorb and hold stray energy information, as crystals are inclined to do.[1] Good stainless-steel pendulums also have a certain amount of "heft," which provides some inertia to help to make measurements more precise.

Dr. Lancaster's advice is given in the context of *closed-system dowsing*. When it comes to *open-system dowsing*, it seems that almost anything goes. When Marty Lucas, a former Board President of the American Society of Dowsers, teaches a basic class in dowsing, he provides students with a short piece of string and a medium-sized nut from the local hardware store.[2] When I received my first pendulum training in the 1980s, I was given string with a lead fishing sinker attached. Such simple and low-cost pendulums

---

[1] Linda Lancaster, "Light Harmonics Institute Summer Intensive," Santa Fe, Aug. 12, 2019.

[2] Marty Lucas, Radionics 101 (Dowsing Segment). Sponsored by the U.S. Psychotronics Association, Skokie, IL. July 25, 2019.

work just fine. But, if using a pretty or exotic pendulum motivates you, by all means, get one!

Open and Closed Systems of Pendulum Dowsing

Let me clarify the difference between open and closed dowsing systems. Generally, when we pose questions or seek things through dowsing, we are using an *open system*. Questions might include:

- Where are my car keys?
- Where is the underground water flow?
- Where is the earth energy line?
- Which supplement should I take this morning?
- And other YES/NO questions like those we'll use later.

By contrast, in a *closed system* the dowser does not pose questions. Instead, one uses the pendulum to indicate and measure real energy flows.

The *Hazel Parcells System Analysis* is an example of a closed system. I first became familiar with this method in the summer of 2019 from Dr. Linda Lancaster, who practices in Santa Fe.

While I am not yet adept, there are several aspects of the Parcells method I want to share. Her system recognizes three pendulum movements for evaluating energy flow. We can observe them easily by dowsing the alternating energies emanating from our fingertips. I've summarized them in the following table.

Pendulum movement	Interpretation	Fingers
Clockwise	Positive	Middle, little
Counterclockwise	Negative	Index, ring
Back and forth	Neutral	Thumb

Interpreting Closed-system Pendulum Rotations

In closed dowsing systems, *positive* and *negative* indicate the direction and movement of energy. They are not judgmental terms. *Positive* does not mean "yes." Nor does it imply "good." *Negative* does not mean "no" or "bad." I am choosing, however, to borrow these closed-system pendulum movements and to adapt their interpretation to decidedly judgmental open-system dowsing.

> **About Dr. Hazel Parcells (1889–1996)**
>
> As a young adult, Hazel Parcells was diagnosed with "incurable tuberculosis, a collapsed lung, a hemorrhaging kidney, and an enlarged heart." After eighteen years of conventional medical treatment and grim prospects, she set off to restore her health on her own. She regained her health and vitality and did not pass away until the ripe old age of 106. During her process of self-healing and expanding efforts to help others, Parcells earned doctorate degrees in nutrition and comparative religions, as well as certifications in chiropractic and naturopathy. She created the radiesthetic analysis system that bears her name and invented a psychotronic device called the Thea Lamp.[3]

Training Pendulum Responses

In the previous chapter, I described a method for pendulum training using flashlight batteries. I used it many times to teach open-system dowsing. However, since my introduction to the Parcells method, I've begun to adapt her closed-system pendulum interpretations, using the fingers of one's hand. Once these movements are mastered, I encourage students to begin interpreting positive movements as YES and negative ones as NO.

To begin dowsing, grasp the pendulum by the pivot bob at the top end of the suspension chain or string. If it lacks a pivot bob, hold the chain or string itself. Use three fingers—the thumb, index, and middle fingers. These three fingers combine neutral, negative, and positive energy flows, and thereby minimize electrical and magnetic influences from the dowser's body.[4]

---

3 Kaayla T. Daniel, "The Legacy of Dr. Hazel Parcels" (sic), Weston A. Price (https://www.westonaprice.org/health-topics/nutrition-greats/the-legacy-of-dr-hazel-parcels).

4 Linda Lancaster, "Light Harmonics Institute Summer Intensive," Santa Fe, Aug. 12, 2019.

Suspend the pendulum above the tip of the forefinger of your other hand. It should begin to rotate counterclockwise. (If the pendulum is not moving, it is okay to nudge it a bit. The goal is to train your body.) Replace your forefinger with your middle finger and observe that the pendulum will wobble some, but soon reverses itself and rotated clockwise. Check your remaining fingers and continue to practice. When you get to your thumb, the motion should change from rotation to a back-and-forth movement to reflect neutrality.[5]

Once you achieve consistent responses you can begin dowsing questions that require YES or NO answers. Remember, you now intend for positive/clockwise rotations to mean YES, and negative, counterclockwise rotations to mean NO. You should interpret neutral, back-and-forth swings to mean "maybe," "not now," or "can't answer."

This is not the only way to learn pendulum dowsing. If you have another method that you prefer, you needn't change.

YES/NO Dowsing Freehand

To train one's self in YES/NO dowsing, begin by posing simple questions with indisputable yes or no answers. For example:

- Am I wearing a green shirt this morning?
- Am I wearing glasses at this moment?
- Is today Tuesday?
- Is my dog named Mister Piddles?

Suspend the pendulum in front of you and ask yourself one of these questions. You can do this in your head or ask it out loud if that helps. If the correct answer is YES, your pendulum will rotate clockwise; if NO, it will circle counterclockwise. As suggested earlier, nudge the pendulum in the correct direction if it is stubborn;

---

5   Ibid.

it is ok to do so when you are in training. Dowse a few additional questions like these to build your confidence.

Asking questions with obvious answers is boring, but training is a matter of learning to crawl before walking. Your objective is the skill to dowse for things that are difficult or impossible to learn by other means. Some examples:

- Is taking this multivitamin now beneficial for me?
- Is red the best color for me to wear today?
- Should I water my ficus plant this evening?
- Should I use bone meal to fertilize my tulips this fall?
- Should I apply aglime to this field before the next crop?

Occasionally we dowse and fail to get a clear yes or no. In other words, the pendulum swings back and forth instead of rotating. This can happen when the question is vague or incomplete. People often fail to specify timing. Note that all of the example questions I provided so far indicate a time period. Red might be an excellent color for you to wear today, but blue might be better tomorrow. And, of course, you should water your houseplant. It will die if you don't. But should you water it *this* evening? You don't want to drown the poor thing!

People might fail to get a clear YES or NO, because it is something they aren't allowed to know—something they shouldn't be asking about. Perhaps it is not within their stewardship or they are not spiritually developed enough. I touched on this in the previous chapter.

How frivolous can one be when dowsing? Is it okay to ask silly questions or to do parlor tricks? Do so if you are so inclined. I don't and won't. In my experience, my dowsing only works reliably when I'm seeking guidance I truly need. An example of this occurred over twenty-five years ago.

We had recently moved to northwest Arkansas and discovered that there were few good records available for our new home. Among the missing was a map or description of where the septic tank was located. Exploratory digging with a backhoe would have destroyed a lot of beautiful landscaping, and hand-digging was clearly out of the question. I do not normally dowse for lost objects, water lines, and the like, but I clearly needed to try in this case. On my first attempt, I found the spot directly over the septic access cover, nearly three feet underground. This was certainly "need to know" information that saved me time and a lot of backbreaking work!

YES/NO Dowsing with a Pendulum Chart

Pendulum charts are handy devices and one alternative to freehand dowsing. Traditional charts are two-dimensional diagrams with answers for the pendulum to point to. They are easily customized for special applications, as you will see later. The chart shown below is an amalgamation of features I've seen on several other charts. It is intended specifically for YES/NO dowsing.

Dowsing charts elicit motions from your pendulum that are different from those described before. You will not want it to rotate but to swing back and forth, side to side, or in between to point toward the answers printed on the chart. But don't be surprised when the pendulum moves elliptically at times. It usually occurs while it is trying to pivot toward the correct answer.

Begin by suspending the pendulum over the midpoint of the X-axis—the point from which the lines begin to radiate outward. Then ask the same indisputable questions you posed when freehand training. I often force the pendulum to move back and forth as I begin asking my questions. This initiates movement, and the device quickly starts to pivot toward the correct answer.

YES/NO Dowsing Using a Finger-Stick

Like the subtle neuromuscular movements that cause pendulums to swing, the finger-stick is another form of ideomotor action. Radionic rubbing (or stick) plates are designed to enhance this sensation, but you don't need an instrument to experience and use the finger-stick. You can learn and practice it on any smooth surface, whether wood, metal, glass, plastic, or vinyl. Practitioners describe the "stick" in a number of ways. Some refer to the feeling one gets when touching sticky tape. Personally, I feel the stick as a dragging sensation—as if my fingertips were impregnated with iron fillings and were suddenly exposed to a small magnet.

Before you start, you might want to apply a dash of cornstarch to the surface to be rubbed so your fingers slide more freely. This is helpful, especially if your hands get sweaty. You don't need much—one-eighth teaspoon is plenty. Ask the same indisputable yes and no questions used earlier while simultaneously rubbing the smooth surface lightly. After each question, follow immediately with the query "Yes?"…"No?" Repeat "Yes"…"No?" questioning until your finger sticks on the correct answer.

Later, as you move beyond training and begin asking "serious" questions, the absence of a finger-stick is understood to mean "maybe," "not now," or "can't answer."

Learning the finger-stick can be more challenging than the pendulum, but it is an exceptionally valuable technique—one you can use unobtrusively in public, where pendulums or other tools might bring unwanted attention. The finger-stick is also the preferred dowsing response for several radionics systems and prepares you should you seek to learn it.

Psychotronics and a Biodynamic Garden

YES/NO Dowsing with a Radionics Instrument

Once you master either pendulum or finger-stick dowsing, you can readily dowse using a radionics instrument. Hieronymus-style analyzers include a rubbing plate that also works with a pendulum; simply suspend it over the plate.

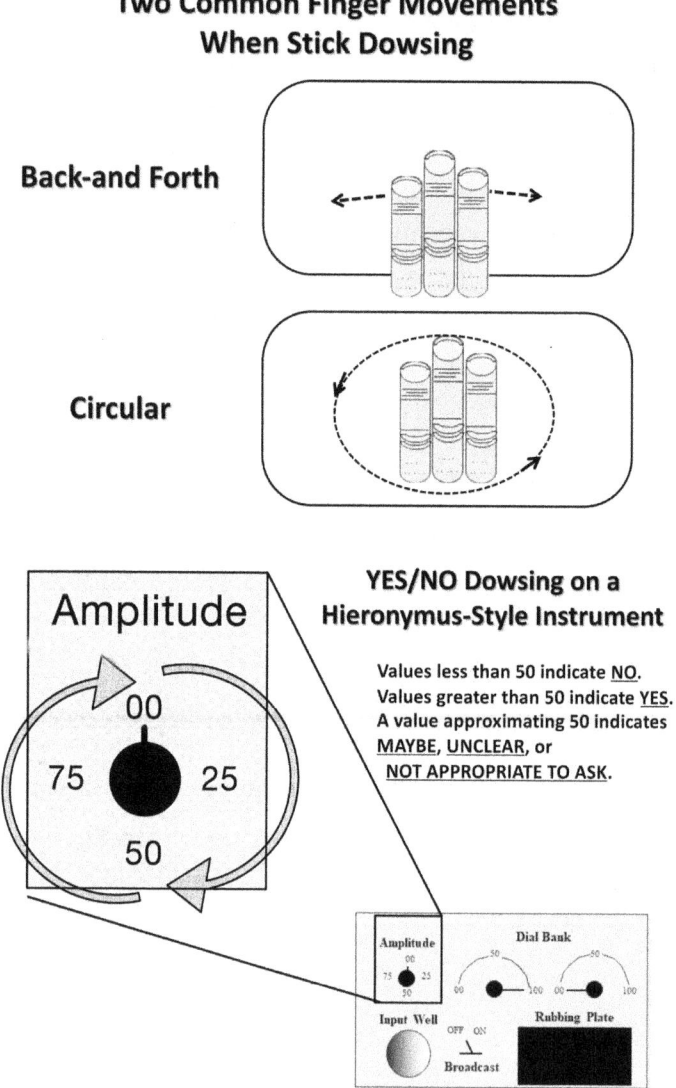

How to Dowse

To begin, set the dial bank to 100–00—the traditional radionics rate for YES/NO dowsing. Ask your question while stroking the rubbing plate or nudging the pendulum in a back-and-forth motion. Simultaneously, rotate the amplitude dial. Stop dial-rotation when you either feel a firm "stick" or the pendulum changes its motion. If the amplitude reads less than fifty, the answer to your question is NO; if greater than fifty, it is YES. A reading of fifty typically means "maybe," "unclear," or "not appropriate to ask."

YES/NO Dowsing with Radionics Template

Radionics Templates

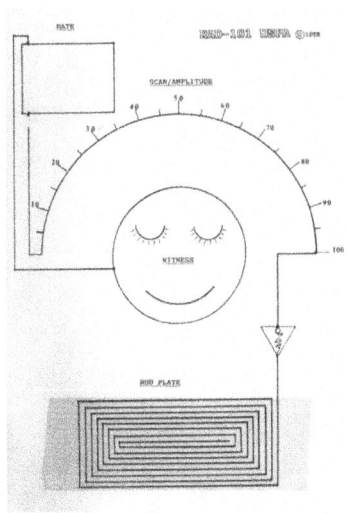 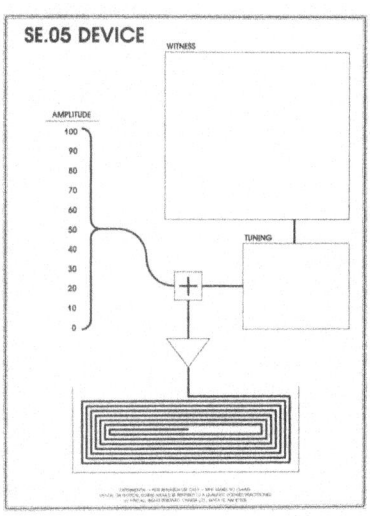

Templates are two-dimensional printed paper or card-stock devices that feature the principal elements of a tangible, three-dimensional radionics instrument. I've seen several of these; two are depicted in the graphic that follows. RAD-101 is copyrighted by the US Psychotronics Association. It is not being produced or marketed at this time. The SE-.05 is also copyrighted, but is commercially available from Little Farm Research.[6]

Templates feature an array of numbers, from zero to ten or zero to one hundred, in lieu of an amplitude dial. Instead of turning a dial, you finger-scan the numbers (see graphic) while stroking or dangling your pendulum over the plate–antenna space at the bottom of the device. As with a three-dimensional radionics instrument, if you get a reading below fifty—or five, if the template uses a lower scale—the answer to your question is NO; if greater than fifty, it is YES. A reading at fifty usually means "maybe," "unclear," or "not appropriate to ask."

Finger-Scanning on a Radionics Template

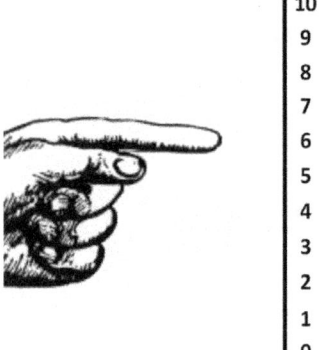

Introducing Scan-dowsing

In YES/NO dowsing, we seek the simplest of answers—either yes or no. When we face more complex situations involving measurement or choosing from multiple options, another method called *scan-dowsing* (also known as *list dowsing*) can be used.

Scan-dowsing with a Pendulum

Scan-dowsing allows us to review a large number of items or options to quickly find what we seek. I will illustrate using an example from gardening. Suppose you plan to grow tomatoes this coming year. You asked neighborhood gardeners and your county agent for recommendations, and you now have a list of ten

---

6 Little Farm Research/Wise Woman Ventures (Lutie Larsen), 993 West, 1800 North, Pleasant Grove, UT 84062.

cultivars.[7] All the suggestions are well informed, but you have neither the space nor the inclination to test-plant all of them. Scan-dowsing is an excellent way to learn which one(s) best suit your conditions and taste.

Place your list in front of you. Hold your pendulum in one hand and nudge it into a back-and-forth motion. Move the finger of your other hand down the list of names, seeking "the optimal tomato varieties to plant in your garden this year." When the pendulum switches to a circular rotation, you've identified one variety that you ought to grow. Continue down the list... you might find another one or two (see graphic).

Scan-Dowsing a List of Tomato Varieties

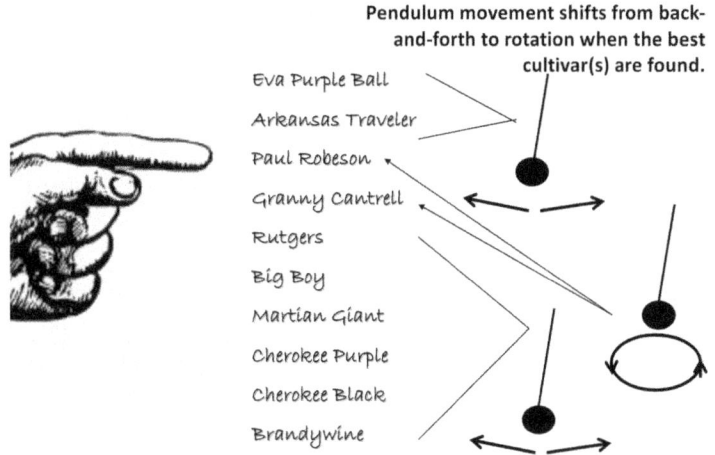

Pendulum movement shifts from back-and-forth to rotation when the best cultivar(s) are found.

Eva Purple Ball
Arkansas Traveler
Paul Robeson
Granny Cantrell
Rutgers
Big Boy
Martian Giant
Cherokee Purple
Cherokee Black
Brandywine

Scan-dowsing Using a Finger-Stick

Scan-dowsing using a finger-stick is much the same as using the pendulum. As advised for YES/NO dowsing, rub a smooth surface lightly while finger-scanning the list of varieties seeking "the optimal tomato varieties to plant in your garden this year."

---

7   The terms *variety* and *cultivar* mean largely the same thing. I use the words interchangeably in this book.

Finger-Stick Dowsing Tomato Varieties

Stroke a smooth surface with one hand while finger-scanning the list with the other. A finger-stick is felt when a desired cultivar is indicated.

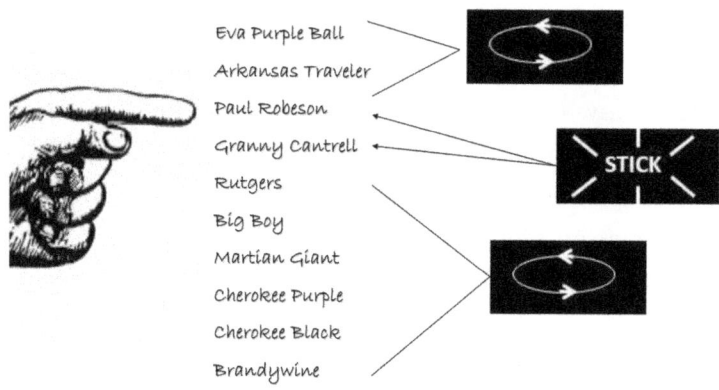

Scan-dowsing with a Radionics Instrument or Template

Scan-dowsing with an instrument or template is virtually the same as pendulum or stick-dowsing. The difference is that instruments and templates provide you with specific places to either suspend your pendulum or rub your fingertips. Set the instrument dial-bank to 00–100—the radionics rate to use when scanning. If using a template, you might wish to write 00–100 on a sticky note and place it in the printed location designated for rates.

Scan-dowsing Number Values

This discussion might appear redundant as it starts as an extension of the list-scanning process we just covered. As you proceed, however, the reason will become clear, especially in the following chapter. Begin by studying the next two graphics. You will note that lists of numbers have replaced the names of tomato varieties.

There are many reasons to dowse for numbers. For example, I might scan-dowse, "How many tomato plants in total should I

Scan-Dowsing a List of Numbers with a Pendulum

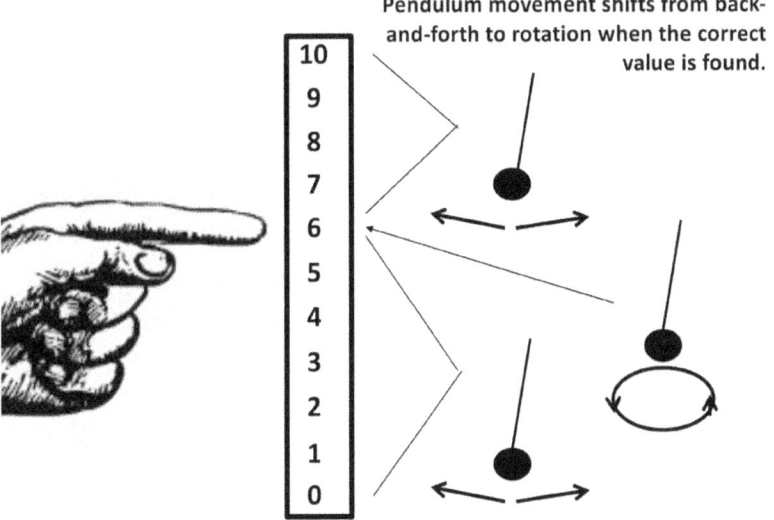

Finger-Stick Dowsing a List of Numbers

Stroke a smooth surface with one hand while finger-scanning the list with the other. A finger-stick is felt when the correct number value is found.

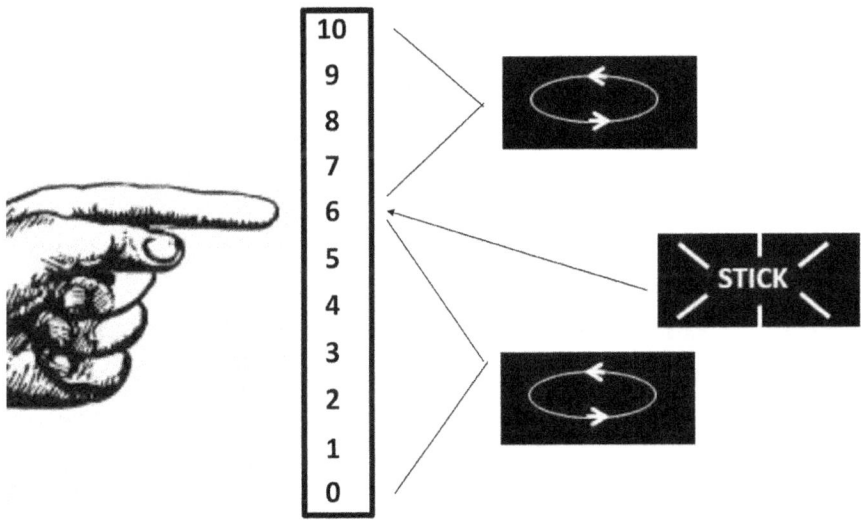

Psychotronics and a Biodynamic Garden

grow in the garden this season?" I can also scan-dowse to determine how many tablespoons of fertilizer to mix in each gallon of water I'll use to drench those tomatoes. In many instances, a written number list is not necessary. You can count in your head or, if you prefer, out loud. It is especially easy when you're seeking whole numbers.

At this point, I will introduce a scan-dowsing application based on the culturally familiar "scale of one to ten" rating system. To illustrate, I'll continue with the garden tomato example. Using the now-familiar list, I scan-dowse each variety, asking, "On a scale of one to ten, how well will this variety perform in my garden this coming season?" As I dowse each value, I record it (see the graphic).

You will note, the most promising varieties are Granny Cantrell and Paul Robeson, though Martian Giant also does well. I will be certain to select among them when I choose what to grow. In fact, I'll probably plant all three varieties in 2020.

Arkansas traveler	6.0	Martian Giant	8.5
Paul Robeson	9.0	Cherokee Purple	7.0
Granny Cantrell	9.5	Cherokee Black	6.0
Rutgers	3.0	Brandywine	4.0
Big Boy	2.5		

But how about a year from now, when I'm trying to decide on the best varieties for the 2021 season? I expect I will review the complete list of ten cultivars once again. I will do so because, even though Rutgers and Big Boy dowsed poorly for 2020, every year is different. Who knows, they might dowse much better for 2021 or 2022. Only when a variety consistently falls toward the bottom several years running would I drop it from my list. And needless to say, I will add to my list any time I learn of new and promising cultivars.

Using a Pendulum Chart to Dowse Number Values

It is very easy to use a pendulum chart to dowse number values. The procedure is largely the same as that described earlier, though the chart is different. This one features numbers from one to ten (see graphic). Suspend your pendulum over the midpoint of the X-axis and nudge it into a back-and-forth movement while asking your question. The pendulum will gradually pivot to indicate the correct value.

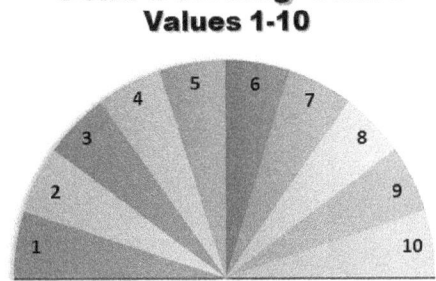

Scan-dowsing Values Using a Radionics Instrument

When scan-dowsing number values on a Hieronymus-style instrument, set the dial bank to 00–100. Pose your question while stroking the rubbing plate and simultaneously rotating the amplitude dial. You have reached the correct number value when a firm "stick" is felt.

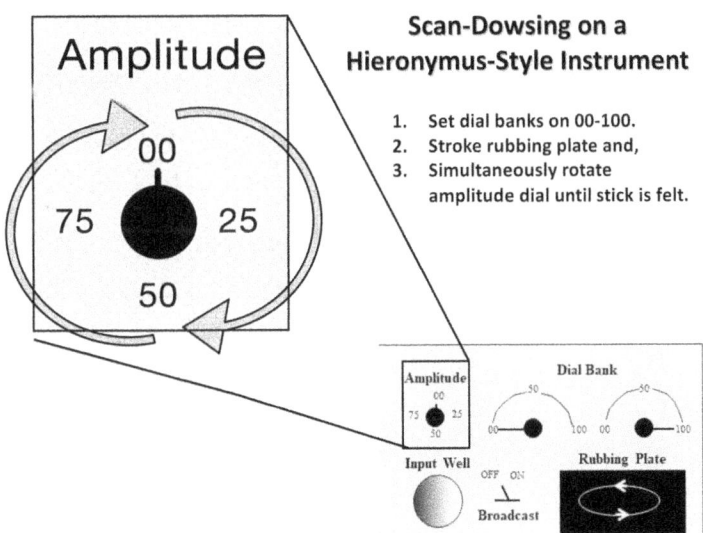

If you are using a pendulum, suspend it over the rubbing plate and nudge it back and forth while rotating the amplitude dial. The correct number is found when the pendulum motion changes.

3.

Establishing Projects and Priorities

In the opening chapter, I described my earliest experiences with radionics. Those were in a real-farm environment in southeastern Oklahoma. That farm was owned by the Kerr Center for Sustainable Agriculture. What I did not mention was that I left the farm project and Kerr Center around 1990, but returned about fifteen years later to develop a new project. The needs and site location were different and I discovered new ways to apply psychotronics.

Designing a New Horticulture Project at the Kerr Center

When I rejoined the Kerr Center, I became responsible for the development of an organic horticulture program. We carved a demonstration site from some of the foundation's pastureland, and called it The Cannon Horticulture Project—named for a family that had once owned the property.

Throughout the next nine or ten years, I used only a few of the same radionics techniques I'd employed about fifteen years earlier on the U-pick operation. The Cannon Project was different. To meet its educational mission, it had to deal with the same conditions as local growers and could not resort to advantages provided through esoteric means. Had I spoken openly about such methods, few, if any, of Kerr's clientele would have used, or even considered, using them. Furthermore, I'd have been fired! Clearly, it would be more honest and helpful to growers if the Cannon Project confronted the same conditions that local growers faced.

Therefore, I explored alternatives for using psychotronics—alternatives that would not compromise the integrity of the Project. I began to think of how I might use it to shape the overall project: prioritizing activities, and selecting better techniques, crops, and plant varieties. In other words, I sought to use psychotronics to increase The Cannon Project's effectiveness for demonstration and education.

I began by brainstorming to identify practical research and teaching demonstrations on topics that would be relevant and informative to growers. These also needed to reflect the Kerr Center's commitment to advancing sustainable farming and gardening methods. Numerous lists emerged, with a multitude of possibilities. A realistic summary list from that time is shown in the following graphic.

A Short List of Focus Options for the Cannon Horticulture Project	
Demonstration vegetable field trials	Integrating poultry into gardening
Heirloom crop varieties	Sorghum syrup production
Organic system design	Small-scale farm technology
Beneficial insect habitats	Plasticulture
Organic compliance and certification	Exotic grain and root crops
Perennial fruits	

I order to find the most worthwhile options—to "winnow the wheat from the chaff"—I adopted the scan-dowsing technique of scoring topics on a zero-to-ten scale, described in the previous chapter. The following table presents a realistic summary of the readings I found for these topics.

On a scale of 1 to 10, how important is each to Cannon Project success for the next 5 years?	
Demonstration-vegetable field trials	8.5
Heirloom crop varieties	8.0
Organic system design	9.5
Beneficial insect habitats	5.5
Organic compliance and certification	4.0

Perennial fruits	3.0
Integrating poultry into gardening	1.5
Sorghum syrup production	1.0
Small-scale farm technology	8.0
Plasticulture	2.0
Exotic grain and root crops	2.5

The most promising options were the four activities with the highest values:

Organic system design	9.5
Demonstration vegetable field trials	8.5
Heirloom crop varieties	8.0
Small-scale farm technology	8.0

I did, indeed, focus on these topics and activities, and have no regrets. The Cannon Project hosted many workshops, tours, and other visitors. Our work there led to the creation of many publications that still inform farmers, gardeners, and the people who serve them. These reports and how-to publications remain available from the Kerr Center for Sustainable Agriculture.[1]

From my standpoint, the most successful effort was in the realm of organic system design. We designed a bio-*ex*tensive management plan that successfully addressed many soil fertility, disease control, and insect pest management issues. Most importantly, however, it demonstrated a non-herbicidal method for suppressing Bermuda grass (*Cynodon dactylon*)—the bane of organic horticulture in the South.

The Shoreline Way Project

As I approached retirement from the Kerr Center, I wondered whether I might continue advancing sustainable agriculture using psychotronics. A world of options seemed to open up for

---

[1] Kerr Center for Sustainable Agriculture: 24456 Kerr Rd, Poteau, OK 74953; Tel: (918) 647-9123; website, www.kerrcenter.com.

Psychotronics and a Biodynamic Garden

me—opportunities to garden, write, volunteer, study, and relax. To help me sort through them I used the same scan-dowsing procedure I'd employed when designing the Cannon Project. I worked on this for months and years before retiring, so many short lists were drafted, dowsed, consolidated, re-dowsed, and recycled. The following is an amalgamation of those early short lists, along with a representative summary of the values I scan-dowsed.

On a scale of 1 to 10, how valuable is each toward my goal of contributing in semi-retirement?	
Operating a small commercial market garden	3.0
Growing food for family, neighbors, and food pantries	8.0
Investigating and implementing biodynamics	9.0
Certifying my land as organic	1.5
Volunteering with the US Psychotronics Association	6.0
Volunteering with the American Society of Dowsers	4.0
Offering more training in radionics	7.5
Master gardener training and volunteering	1.0
Volunteering with the Botanical Society of the Ozarks	1.5
Writing another book on subtle energy and agriculture	9.0
Indolence and debauchery	0.5
Look for another option	1.5

My present efforts to join biodynamics and psychotronics together in my garden and write about it grew out of these dowsing exercises. They are at the center of what I call my Shoreline Way Project—named for the road on which we live.

There are many ways we can use dowsing to add direction to our lives. This is one way that I've chosen to do it.

4
Plant–Earth Alignment

This chapter departs somewhat from the path of discussion so far but, as another example of dowsing, it fits well here. Besides, the concept of *critical rotation position* has been a subject of psychotronics research and application for a long time and deserves mention in this book.

The Critical Rotation Position

Critical rotation position (CRP) refers to the optimal orientation of a plant's stem and roots to the Earth's energy field. When plants are growing in their critical rotation position, they are in *perfect resonance* with the life forces that sustain them and are "receiving the optimum quantity of vital radiations."[1]

When we sow plants directly into the garden, they germinate and grow naturally into their critical rotation position. However, when we seed them in pots or flats for transplanting, CRP is often compromised as we shift them about. This does *not* mean that CRP-compromised plants will die; it *does* mean they experience additional stress and may not thrive as well as they might.

---

1   Day, *New Worlds beyond the Atom*, pp. 80–81.

Scan-dowsing for Critical Rotation Position

It is easy to locate the CRP for a houseplant. Using a pendulum, or finger-stick on a smooth surface, rotate the pot until you get a firm dowsing response.

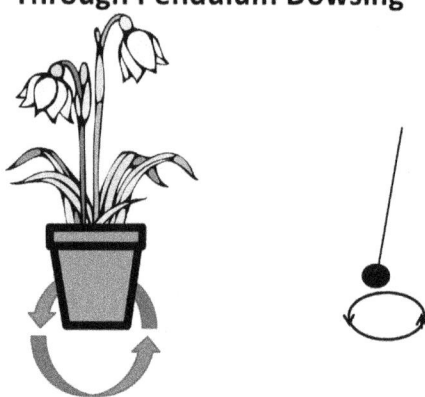

Finding the Critical Rotation Position (CRP) Through Pendulum Dowsing

Orienting garden transplants is a bit more challenging. When I'm in the garden and ready to set a transplant, I rotate it in my free hand while observing my pendulum. When I locate its CRP, I'm careful to maintain the seedling's orientation as I place it in the ground. If I were transplanting commercial seedlings from four- or six-packs, I dowse the whole pack as a unit to find the correct orientation.

Because I've long grown my own transplants for both the Cannon and the Shoreline Way Projects, I found a more expeditious way to preserve their CRP. When seeding my trays, I mark them on one side with the cardinal direction they face. Thereafter, should I move them into the greenhouse or place them outside for "hardening,"[2] I can correctly orient the plants. Later, when transplanting, I keep

---

2 *Hardening* is the process of exposing transplants (seedlings) gradually to outdoor conditions. It enables your transplants to withstand the changes in environmental conditions they will face when planted outside in the garden (University of Maryland Cooperative Extension, https://extension.umd.edu/hgic/hardening-vegetable-seedlings).

the trays so-oriented and set the seedlings accordingly. If you grow commercially, none of these options are probably doable. It might make more sense to accept the bit of added stress and, perhaps, compensate your plants through other means.

Late this past winter, I made an investigation into CRP. I wondered if germinating plants immediately locked into their critical rotation position, or whether it became established later in the growth cycle. I created a rate for "plant establishment of critical rotation position (CRP) during and after germination" (35.20–64.45). Using just-planted broccoli and cabbage seeds, I compared the readings to general vitality (09.00–49.00). Under good germination conditions in high-quality transplant mix, the broccoli seedlings and one of three cabbage varieties reached a CRP value of about one-third general vitality in six days following planting. The other two cabbage varieties lagged behind. It appears that seedlings reaching one-third of their final CRP are not yet locked into it and won't experience stress if oriented in another direction.

I attempted this preliminary experiment because many growers start transplants by scattering seeds over a tray filled with a low-cost germination mix. Shortly after emergence, seedlings are plucked out and replanted into multicell planting trays or peat pots filled with a higher-cost but more nutritious growth mix. The process saves money and greenhouse space.

My preliminary finding suggests that this two-step method for growing transplants might not compromise critical rotation position if the first transplanting is done early enough. But since the time window between planting and one-third CRP is quite narrow, it may not be practical. Obviously, mine was not a well-planned experiment and much more study is warranted before good conclusions can be drawn.

Psychotronics and a Biodynamic Garden

Finding the Front Door

There is a notion that plants—especially trees—have *energy doors* and, specifically, a *front door*. Dowsing a tree's front door probably has numerous purposes, but I am most familiar with its use for proper placing of French or Lakhovsky coils around tree trunks. Made of copper wire, these coils are used for healing or to enhance tree growth. It is not a practice I've made use of, but Australian dowser Alanna Moore has spent many years working with these devices. She writes that the front door is part of the plant's aura and the spot where concentrated flows of energy can be detected. She suggests that it might function much like a heart chakra or a seat of consciousness for the plant.[3]

According to another experienced dowser, Harvey Lisle, the author of *The Enlivened Rock Powders*, "…the energy door of a tree is associated with the neutral point of a magnet—located where the positively charged atmosphere meets the negatively charged Earth."[4] Lisle's description appears to correspond with the location identified by Lutie Larsen for the *physical support chakra*—one of four major plant *energy centers* she has found in her research.[5]

A Contribution from T. Galen Hieronymus

In *Cosmiculture*, T. Galen Hieronymus describes one of his experiments that demonstrates the importance of Earth's magnetic alignment to the vitality of plants.[6] The demonstration used a basic radionics analyzer, a small potted plant, and a diagram similar to the one in the following graphic, properly oriented to the cardinal compass points.

---

3   Moore, *Stone Age Farming: Eco-agriculture for the 21st Century*, pp. 71–72.

4   Ibid., p. 181.

5   Larsen, *Little Farm Tips and Techniques for Farmers*, pp. 134–135.

6   Hieronymus, *Cosmiculture*, pp. 8–9.

Diagram for Hieronymus' Plant Orientation Experiment & Radionic Readings

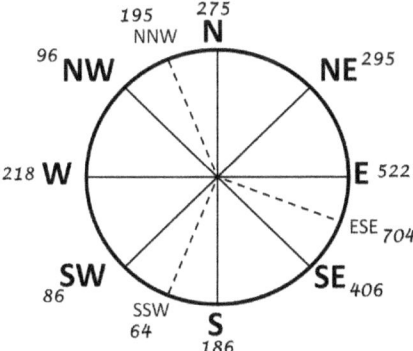

After marking a reference point on the container, Hieronymus rotated the plant through the cardinal points, taking measurements of general vitality (rate: 09–49). The readings he recorded are also shown on the graphic. They suggest that the Critical Rotation Position for the plant was to the east-southeast. The least beneficial orientation was to the south-southwest.

5

Intent and Manifestation

Radionics programming is a psychotronic procedure for broadcasting one's intentions into the universal energy field. You might use it to support your garden or another project. You might use it to help manifest something you or others need, such as a good job, improved health, or a good rain in the middle of a drought. You might use it as a prayer for peace and justice.

The renowned radionics teacher Steve Westin (1944–2015) taught me this technique in 1986. I've been well-rewarded by this procedure and treat it with great respect. I won't use it frivolously. Radionics programming, which I prefer to call *psychotronics programming*, offers us the power to set things in motion that may have profound effects. We don't want to do anything we might regret later. Those catastrophes, large and small, happen when we broadcast wishes and desires based on selfishness, vindictiveness, caprice, half-baked ideas, and the manipulation of others. Even if your heart is pure, problems can still result from failure to think things through.

Developing an Intent Statement

Clarity of intent is vitally important in psychotronics. It is critical for obtaining valid information; it is even more so when attempting to shape reality through broadcasting. Shaping reality *is* what we try to do when we project intentions out into the universal energy field. At the risk of sounding pompous, I've learned to "expect" results from psychotronics programming. I suspect my successes have much to do with the time and attention I give to creating my intent statement through dowsing. I focus not only on getting a clear statement, but on the "rightness" of the outcome(s). If what I'm wanting does not support health, justice, and harmony, then I really *don't* want it.

For my examples, I will focus on gardens and gardening projects. My first example of an intent statement logically follows from the previous chapter's discussion of project planning. There I wrote about how to identify priorities. Once you have that knowledge in hand, you can proceed to psychotronic programming, in which you broadcast your intentions—in the form of an intent statement—to help make things happen as you wish.

When you write an intent statement, avoid ambiguous language. Your key words *must* have clear meaning to you. In the following sample intent statement (adapted from my own work), I use the terms *organic* and *psychotronics*. Neither term might be clear to the average person, but they certainly are to me and, I'm trusting, are clear to you, as well.

Example of an Intent Statement

> Manifest, beginning Summer 2016 and extending through the growing season, a food garden at Shoreline Way, Fayetteville, Arkansas, using organic cultural practices and psychotronic methods. Ensure that this garden and associated gardening activities produce:
> - Abundant Quality Food
> - Enhancement of the soil and the environment
> - Experiential learning to support writing a helpful book within ten years
>
> But not my will but thine be done.
> No Harm to Anyone
> George L. Kuepper

While it's no masterpiece of prose, this intent statement contains several important features:

- A timeline: *summer 2016*... clarifies the season with which I'm concerned. In *my* circumstances, this happens to be important. It might not be for all intent statements. One should dowse to determine whether timelines are needed for all or part(s) of your statement.
- A precise location: *Shoreline Way, Fayetteville, Arkansas.*
- Command words: The words *manifest* and *ensure* strengthen intent. I've provided a list of additional command words you might consider.
- *But not my will but thine be done*: This short declaration acknowledges that you want something but recognize that the Universe might have other plans. If you don't like the Judeo-Christian wording but agree with the sentiment, choose words and wording that makes you comfortable.
- *No harm to anyone*: This is another customary and recommended declaration. We don't want our desires or needs fulfilled through tragedy or someone else's misfortune.
- Signature: You should not be hesitant to sign your name to your intent statement; you *do* "own it."
- One added note: Despite my diatribe about clarity, I was intentionally vague about the book in my statement. When I wrote this in early 2016, my conception of what my book might be about was too vague to provide details. All I knew was that it should a priority for my future. I dowsed the frame of ten years, giving me enough time to produce an acceptable result.

When I write an intent statement, I dowse every sentence, and all the key and command words. The result seldom makes great reading, but my intent becomes crystal clear and that is what's important. When I *think* I've completed my statement, I dowse further to see if additional sentences, phrases, or words are needed. It is discouraging when you find that more needs to be written, because it can be so difficult to nail down! I have no specific system for easily finding such omissions. I just dowse around until I hit on the right thing, and I usually do. It's time-consuming, but it is worth persevering.

Additional Command Words			
Activate	Decrease	Heal	Release
Align	Detoxify	Improve	Remove
Assimilate	Develop	Increase	Repair
Balance	Discourage	Inhibit	Restore
Block	Eliminate	Mobilize	Reverse
Cleanse	Encourage	Neutralize	Stabilize
Clear	Energize	Optimize	Stimulate
Complement	Expand	Prevent	Strengthen
Consolidate	Germinate	Purify	Unblock
Correct	Harmonize	Reduce	

Adapted from Lutie Larson, "What Are Command Words?" *The Radionic Homestead Report*, vol. 7, no 4, Jul./Aug., p. 5.

Basic Intent Statements for Gardening

Intent statements need not be as complex or widely encompassing as in the previous example. Examples of simpler intent statements follow:

Examples of Intent Statements

> Protect this Garden from early frost and freezing conditions until November 1, 20__
> But not my will but thine be done.
> No Harm to Anyone
> *Signature*

> Detoxify and Cleanse this field of all herbicide residues and their chemical break-down products.
> But not my will but thine be done.
> No Harm to Anyone
> *Signature*

These examples do not specify which garden and field we are interested in. If you formulate an intent statement in this way, you will need to include a witness of the targeted location later, when broadcasting.

Examples of a few more intent statements follow:

More Examples of Intent Statements

> Stimulate the beneficial biology and enhance natural breakdown and transformation processes in all compost piles constructed on Cold Comfort Farm.
> But not my will but thine be done.
> No Harm to Anyone
> *Signature*

> Enhance my skills and abilities for supersensible perception of the natural world by and through gardening.
> But not my will but thine be done.
> No Harm to Anyone
> *Signature*

Gut Gardening at Perelandra

Perelandra is the title of a book by C. S. Lewis (1898–1963), in which he describes a fantasy world on the planet Venus. It is also the name adopted by Machaelle Small Wright for her spiritual center in Virginia. At Perelandra, Machaelle does research and education about interacting with nature intelligences—devas and nature spirits. That work revolves mainly around gardening.

Small Wright teaches several levels or methods for spiritual gardening. The most basic level is *gut gardening*—a simple approach to working with devas and nature spirits. It requires making an intent statement as well. Machaelle calls it a *DDP statement* because it clearly states *definition, direction,* and *purpose*. Here is one example from her workbook: "A kitchen garden with vegetables and herbs that can provide fresh produce through the late spring, summer and early fall for my family of four (two adults and two children ages eight and twelve)."[1]

DDP guidelines are excellent and might assist you as you develop intent statements.

From Intent Statement to Intent Card

Once you've written a satisfactory intent statement, you should create an *intent card* (or *witness card*) for psychotronics broadcasting. The manner in which you make the intent card is important as it adds further strength and clarity to your intentions. Begin by finding a clean desk or tabletop on which to work.

For paper, I recommend clean, white unlined paper or card stock. Three-by-five-inch unlined index cards are more than adequate. Actually, though, they are a bit large, especially if you plan to use a radionics instrument for broadcasting, so cut them smaller if you need to. You might consider purchasing a small pack of index cards and keep them hidden so others will not handle and compromise them. Do not leave fingerprints on your intent card or otherwise contaminate them. I recommend using fresh nitrile or plastic gloves throughout the process. Sometimes I also use tweezers—the kind preferred by stamp collectors seem to work best, as their tips are broad, thin, and flat.

I recommend you handwrite your statement on the card. Use either a graphite pencil or a pen with India ink. Both have paramagnetic properties[2] that aid in holding intent.

Once you have written your intent statement and signed it, the intent card is complete and can be used for broadcasting. However, you must continue to take care when handling it. It is still vulnerable to contamination and compromise. I suggest you either laminate or seal the intent card in a glassine envelope, especially if you

---

1 Small Wright, *The Perelandra Garden Workbook*, pp. 21–22.

2 *Paramagnetism* refers to a property of certain materials that are weakly attracted to magnetic fields.

hope to reuse it. Glassine and laminate are both considered neutral and safe when used in a broadcast; either can be easily wiped clean of dust and fingerprints; soft toweling dampened with distilled water is recommended. Be especially careful if you laminate. It is much too easy to trap finger prints under the lamination and ruin the witness card.

Rate Scanning

If you have a radionics instrument, or a radionics template, there is one additional step I encourage before laminating, sealing your card in glassine, or broadcasting. It will enhance your process further if you create a radionics rate for your intent and add this to the intent card.

There are a large number of radionics rates currently in use. You can find them in specialized rate books or catalogs, as electronic databases, and organized in analysis sheets. While we are encouraged to use published rates, sometimes you can't find one that fits your specific needs. This is certainly true when you want one that captures your customized intent statement. In such circumstances, we must develop our own radionics rates. This process is often called *rate scanning*. Do *not* confuse it with scan-dowsing. They are quite different. And note: *all* radionics rates are developed using some variation of the technique I am about to describe.

The graphic on page 69 depicts rate-scanning on a Hieronymus-style instrument. To find the optimal rate for the intention on a witness card, begin with amplitude and rate bank dials set to zero, and then slip the uncontaminated intent card into the input well.

Finger-stroke the rubbing plate or nudge your pendulum into a back-and-forth swing while simultaneously rotating the left-hand rate dial (*not* the amplitude dial). When a firm stick is felt or a change in pendulum motion is observed, stop rotating the dial and leave it set to the number indicated. Repeat the procedure with the

Intent and Manifestation

Scanning Procedure:
- the witness card is in the input well
- stroke the plate while rotating the left-hand dial until stick
- stroke the plate while rotating the right-hand dial until stick
- resulting numbers (xx.xx—xx.xx) are the rate for the intention

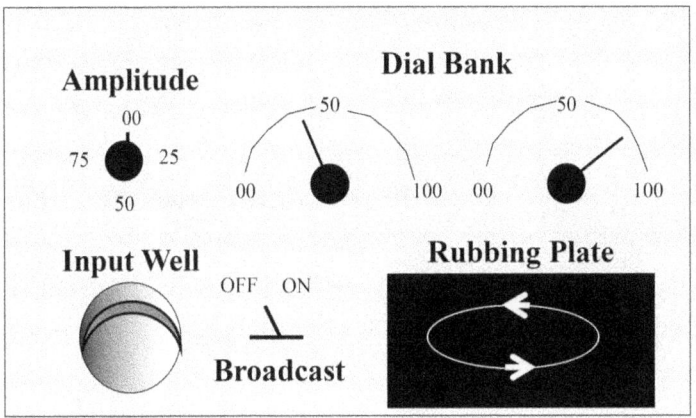

right-hand dial. The number combination indicated by the rate dials is your newly scanned radionics rate.

For the sake of our example, let's assume the numbers you scanned were 38–80.5. I always take additional time to dowse and fine-tune the rate; the more precise it is, the more power it has. I will not take the time here to walk through that exercise; it can be tedious.

As a last step, I will YES/NO dowse to ask if the rate I developed is truly suited for broadcasting this intent. If "yes," I will record it on the card as shown in the graphic. However, since the new rate *is* the access code for this specific intent, I *do* have the option of broadcasting the rate by itself, without the intent card.

Option: Adding a Rate to the Witness Card

> Stimulate the beneficial biology and enhance natural breakdown and transformation processes in all compost piles constructed on Cold Comfort Farm.
> But not my will but thine be done.
> No Harm to Anyone
> *Signature*
>
> XX.XX—XX.XX

Scanned Radionic Rate For the Written Intention

6

Psychotronics Broadcasting

Following the process of creating intent cards, we proceed to psychotronic broadcasting. I will continue to focus on using intent cards. Please understand, however, that the same procedures are used when broadcasting reagent energies. You can substitute specimen samples of fertilizers, gemstones, color filters, homeopathic medicines, BD Preparations, essential oils, or almost any other substance for an intent card when broadcasting *if* they are deemed appropriate. Dowsing is your key to finding whether a reagent, an intent card, or simply a radionics rate written on a piece of paper is "safe and advisable to broadcast."

Broadcasting with a Radionics Instrument

Begin by neutralizing the input well, and then insert the intent card. If your intent statement does not clearly state the location or the specific subject you are targeting, you need to add witnesses such as soil samples from the garden or field, seeds of the crop being planted, and/or a good-quality photo. (I address these details in the next chapter.) Set the rate dials to 00–100.

Dowse to confirm that it is "safe and advisable to broadcast." If you receive a NO response, determine whether you've set up the broadcast correctly. Perhaps you need additional witness(es). If your setup checks out, dowse whether the broadcast should be made at a later time. If "no," you have a dilemma. It might be a minor issue or it could be significant. Perhaps you are venturing beyond your stewardship or beyond your capabilities as a practitioner. I know of no quick shortcuts to finding the cause(s) of repeated blockages. If it gets too frustrating trying to dowse the reason(s), I abandon the project. I never, and *you* should never, proceed with any radionics broadcast that is not "safe and advisable."

If you get the go-ahead, however, toggle the broadcast switch "on," and dowse for the time required for the broadcast (see the following graphic).

Broadcasting with a Witness Intent Card:
- witness card in the input well; add subject's witness if needed
- set rate dials on 00-100
- toggle broadcast switch "ON"
- dowse for time using amplitude dial

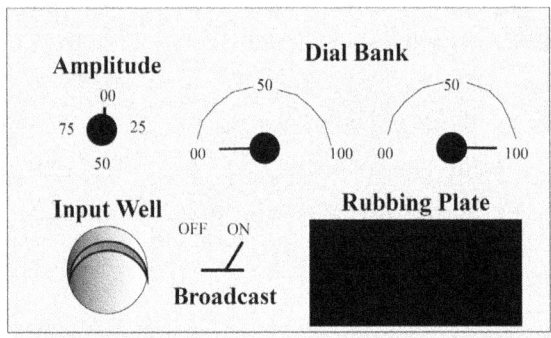

If you scanned a rate for your intent statement, you *can* set the intent card aside, set the rate on the instrument dials, and broadcast as shown in the next graphic. Be sure, though, to include additional witnesses of target locations and subjects, should they be required.

Broadcasting with a Scanned Rate for the Intent:
- subject's witness in input well if needed
- set rate dials on scanned "intent" rate
- toggle broadcast switch "ON"
- dowse for time using amplitude dial

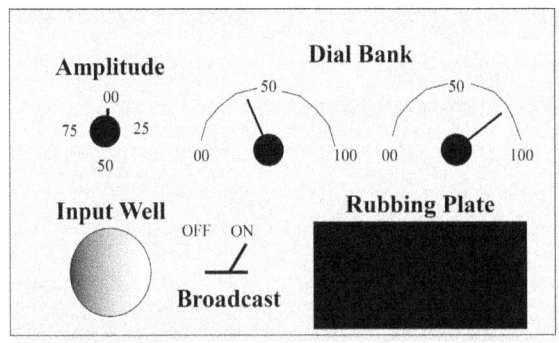

Magnetic Broadcasters

Researchers and practitioners have discovered a wide array of broadcasting devices besides radionics instruments. Many of these feature magnets. The *magnetron* is one of the best-known examples. It was developed in the early 1950s by a two Belgian brothers, Felix and William Servanx.[1] It was subsequently popularized by Christopher Hills and the University of the Trees, Santa Cruz, California.[2]

In *Stone Age Farming*, Alanna Moore recounts a demonstration by Hills in which a magnetron was used to help control sowbugs in a garden. This entailed killing one or more sowbugs and broadcasting their "death vibrations" to the infested plots. Moore also discussed using the magnetron to broadcast love to a garden—certainly a more benign action.[3]

According to Roger Smith:

> The magnetron acts as a powerful radiesthesic link to the earth's magnetic field. Eight magnets of alternating polarity are arranged in a circle on its face and by aligning the device with the earth's north-south magnetic axis we "tune" it.[4]

I acquired two magnetrons in the early 1990s, when they were still commercially available from University of the Trees Press. I use them for sustained broadcasts—one for protection of my homestead, the other to support my vision for the Shoreline Way Project. My intent cards are prepared as I described in the preceding chapter. Each time a broadcast is initiated, I dowse to determine when the witnesses and/or the intent card will need to be "reviewed," and

---

1 Roger Smith, "The Magnetron," in Allen et al., *Energy, Matter, and Form*, pp. 279–280; Servanx Publishers (http://www.servranx.com/QuiSommesNous.aspx).

2 Hartman, *Shamanism for the New Age*, pp. 85–89.

3 Moore, *Stone Age Farming*, pp. 78–79.

4 Roger Smith, "The Magnetron," in Allen et al, *Energy, Matter, and Form*, pp. 279–280.

mark this on my calendar. When the date comes around, I dowse to determine what should be done. Sometimes, cleaning the magnetron's surface is all that's required. Often, though, I need to reword the intention or replace a witness.

One of my magnetrons is shown below. The left-hand view is an unobstructed view of its face. The small white circles within the large black circular band show the location and arrangement of eight small magnets embedded in the wood framing. The right-hand view depicts the same magnetron while actively broadcasting.

Magnetron

This Magnetron is one of two I purchased around 1990. They serve as a psychotronic broadcasters and direction-finders.

The same Magnetron is shown in broadcasting mode. Note that a witness intent card, a photographic subject witness, and a pocket polarizer are positioned in the center of the unit.

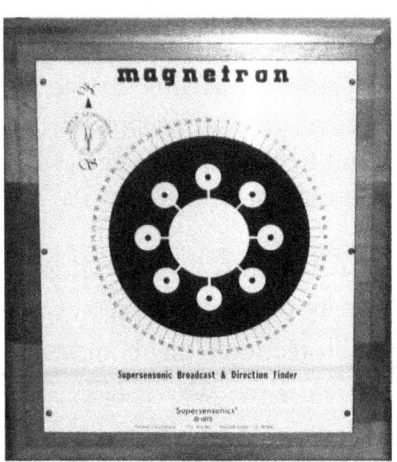

The magnetron should be oriented to the earth's magnetic field, using a compass, and dowsing to fine-tune. It should be placed where it will receive plenty of sunlight.

The intent card is placed in the center and the subject's witness (a photo in this case) is laid on top of it. The object shown on top is a pocket polarizer. I dowsed to determine that it was beneficial to the broadcast.

As I've hinted already, there are many ways to arrange witnesses and enhance magnetron broadcasts. Jane Hartman describes the additional use of colored glass or jell sheets, glass containers of reagents, and using full spectrum lights. However, as she points out, "There are no limits upon your creativity and the use of these methods, except those imposed by the operator."[5]

Magnetic broadcasters can assume other forms. The late Mark Moeller, an Arkansas blueberry grower, dowser, and radionics practitioner, described the "molecular generator" at a workshop he co-taught with Gene Litwiller in the late 1980s.[6] I've reproduced Moeller's drawing below, made from my sketchy notes.

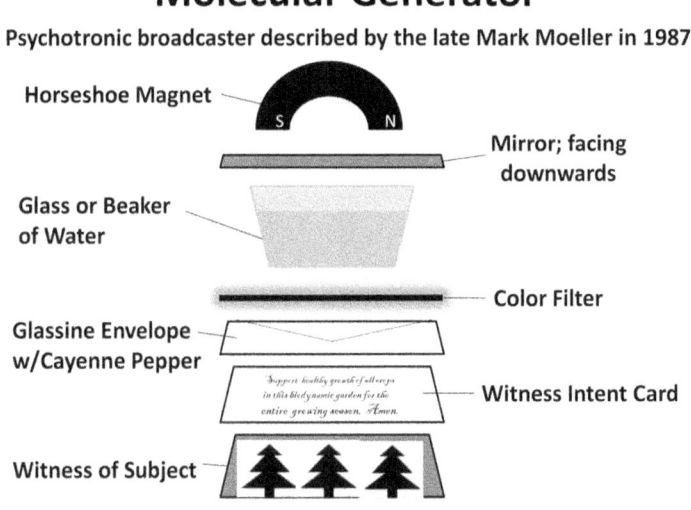

Moeller provided few details about the molecular generator. However, I discovered a brief reference to a similar device in Jane Hartman's book. She references the writer and dowser Gabriela Blackburn, who uses a one-pound horseshoe magnet with a

---

5   Hartman, *Shamanism for the New Age,* pp. 85–89.
6   Mark Moeller and Gene Litwiller, "Agricultural Radionics Workshop," Ramada Inn, North Kansas City, MO, Nov. 28–29, 1987.

fifty-pound pull for her broadcaster. She adds that the north pole of the magnet should always be oriented to the earth's magnetic north.[7]

While I hope the diagrams and recommendations are useful, do not place too much weight on every detail. You can, and should, dowse to determine:

- whether the glassine envelope should contain cayenne, another herb, or if it is even necessary;
- which color filter will work the best;
- whether the water used should be structured, distilled, rainwater, and so on;
- if the container holding the water can be plastic or Styrofoam;
- how powerful the magnet should be.

Symbolic Broadcasters

In *The Incredible Hieronymus Machine*,[8] Joseph Goodavage (1925–1989) recounts numerous observations and comments by John W. Campbell Jr. (1910–1971), a science-fiction writer, editor, and radionics enthusiast, regarding his personal investigation of T. Galen Hieronymus' invention. Among them is a record of Campbell making a paper drawing of the Hieronymus instrument using India ink. In his correspondence with Hieronymus, Campbell insisted that his drawing worked as well as "the real thing."[9] Goodavage's writing about Campbell's findings encouraged many folks to experiment with symbolic devices made from paper and card stock.

Radionic templates, depicted and discussed earlier, are among the best examples of symbolic instruments. They are well-suited for

---

7 Hartman, *Shamanism for the New Age,* pp. 85–89.

8 "The Incredible Hieronymus Machine," in Goodavage, *Magic: Science of the Future,* chap. 4, pp. 50–71.

9 Although I am not referencing or reproducing drawings from the book, readers interested especially in symbolic instruments should investigate G. Harry Stine, *Mind Machines You Can Build: Move Things with Your Mind and Other Experiments.*

dowsing, but can also be used for broadcasting. Practitioners sometimes call them "radiesthetic devices." They require more sustained focus by the operator for broadcast efficacy. How well do they work in specific circumstances? How long does the operator need to stay focused? Dowsing and subsequent observations are needed to answers these questions.

Similar to a template is the Radionics Paper Chart machine, shown in the following graphic. I first learned of this 2-D device from Wynelle Delaney at a 2001 dowsers' conference in Arkansas.[10] During her presentation, Wynelle provided information previously developed by Hank Smyth in 1991 for a similar event in the state.[11]

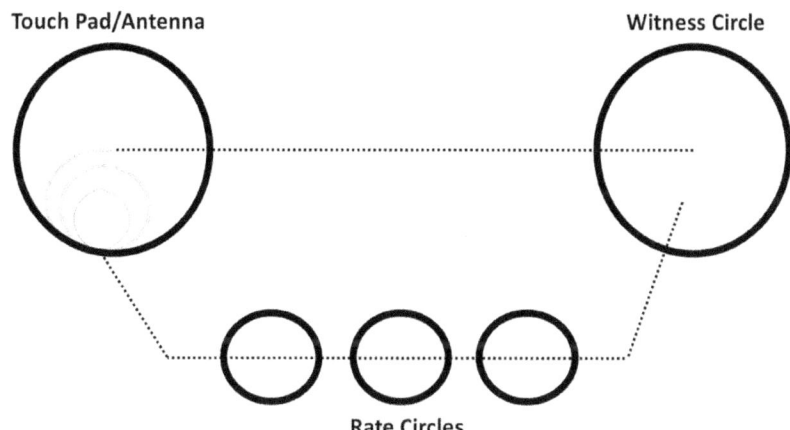

Sources: Smyth, Hank. 1991. Dowsing and Radionics—Keep It Simple. Mid-South Dowsers Conference. Fayetteville, Arkansas. April 19-20. *and* Delaney, Wynelle. 2001. Presentation title unknown. Central ASD Conference. Eureka Springs, Arkansas. April 27-29.

---

10 Wynelle Delaney, Central ASD Conference (presentation title unknown), Eureka Springs, Arkansas. Apr. 27–29, 2001.

11 Hank Smyth, "Dowsing and Radionics: Keep It Simple," Mid-South Dowsers Conference, Fayetteville, Arkansas, April 19–20, 1991.

Psychotronics and a Biodynamic Garden

The device is largely self-explanatory except for the line that links all five circles. It begins in the middle of the witness circle and runs to the center of the touch pad. Here it forms three counterclockwise circles before leaving the pad and connecting the three rate circles. Finally, the line continues back to the witness circle just shy of its starting point. The operator completes the link with a pencil when starting the broadcast. When rates are used, they are composed of three single-digit numbers scan-dowsed from a list of numbers or using a dowsing chart, as shown in an earlier chapter. The rate numbers are written in the circles provided. As different rates are needed, they can be erased and changed but, as Smyth pointed out, this device is intended to be readily disposable. Practitioners can make multiple photocopies and use a fresh one each time.[12]

It is not clear whether Smyth originated this symbolic device or stumbled on it elsewhere. I was pleased to discover something similar in one of Alanna Moore's books. She credits the device to another Australian dowser, Ann Miller. Moore does not describe it as a broadcaster, but details its use as a radionics analyzer for farming and gardening.[13] (If you do so, do the planet a favor and recycle the used paper.)

Another example of a symbolic broadcaster is one used in the Sai Sanjeevini system, "a self-learning, prayer-based spiritual healing system."[14] Sanjeevini has several features and practices in common with contemporary radionics, including potentization and the use of prepared healing cards similar to those used in the Malcolm Rae radionics system.[15]

---

12 Ibid.
13 Moore, *Stone Age Farming*, pp. 62–63.
14 From the Sai Sanjeevini website: www.saisanjeevini.org.
15 Malcom Rae pioneered a system of radionics called Magneto Geometry, which uses prepared cards rather than numbered rates. Young, Sue. 2009. Malcom Rae (1913–1979). Sue Young Histories. May 30 (https://www.sueyounghistories.com/2009-05-30-malcolm-rae-1913-1979).

Example of a Sai Sanjeevini Card

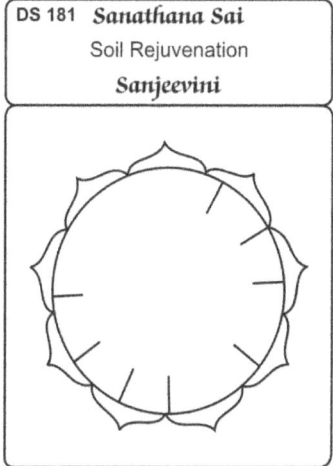

Reproduced with permission from Vinod Nagpal, Sai Sanjeevini Seva. October 10, 2019

Sai Sanjeevini Broadcaster

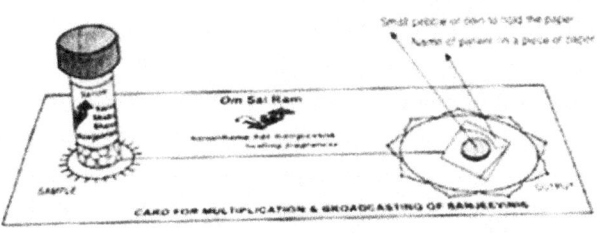

Reproduced with permission from Vinod Nagpal, Sai Sanjeevini Seva. October 10, 2019

The Sai Sanjeevini system employs a symbolic device specifically for remote broadcasting. The broadcaster is depicted on the top left in the preceding graphic; on the bottom-right, it is shown broadcasting a remedy to a subject.

Power Pods

Power pods are small psychotronics broadcasters made by combining two commercially available components[16]:

1. A Betar Dual-Spin Coil, made by Kelly Research Technologies. The coil is a pocket-sized version of the Perfect Spiral found in their phase-array reaction plates and console instruments.[17]
2. A three-inch-diameter tensor ring, made by Twisted Sage (South Dakota). Tensor rings are described as superconductors that neutralize magnetic fields and stabilize the biomagnetic and energy fields of the body.[18]

The Betar coil fits neatly inside of the tensor ring to form the power pod as shown in photos on the next page. On the right, we see the power pod in broadcasting mode, using an intent card and a subject's photographic witness.

Power pods also work for "local" broadcasting. They can radiate an intention, protection pattern, or other positive energies for several feet around the pod. We accomplish this by *not* including a subject witness when broadcasting. So, instead of broadcasting to a distant target, the pod radiates locally. Pete Radatti describes this function in his book *A Fun Short Course in Beginning Radionics*.[19]

---

16 Radatti, "Pennsylvania Pete," *A Fun Short Course in Beginning Radionics*, 3rd ed., Conshohocken, PA, self-published, 2016, pp. 97–98.
17 Kelly Research Technologies (http://kellyresearchtech.com/ka-analyzers.html).
18 Twisted Sage (https://twistedsage.com/what-is-a-tensor-ring).
19 Radatti, pp. 99, 254.

Psychotronic Power Pod

Betar Coil
Tensor Ring
Power Pod in Broadcast Mode
Note intent card and photo witness

The Question of Power and How to Increase it

Personal radionics instruments, such as the Rogers instrument and the small units marketed by Kelly Research Technologies, are considered low-power instruments. (So are most of the alternative broadcasters presented here.) As such, they are not promoted for agricultural broadcasting when the field area exceeds two and a half acres. This makes them ill-suited for large farms, but more than adequate for gardens and most market gardens.

The two-and-a-half-acre limit is a rule of thumb and should be treated as such. Depending on your circumstances, the area you can treat might be smaller, but it might also be larger. Sometimes all that is required is a longer broadcast time, or positioning the unit over an earth energy spot. Dowsing is the only sure way I have for knowing if my instrumentation is powerful enough for a task. You should make that determination any time you question the power of your device for a landscape broadcast.

Two Early Cosmic Pipe Designs

Pictured on the right is a rendering of an original Cosmic Pipe developed by T.G. Hieronymus. Constructed largely of PVC pipe, the angled plate at the top contains a sheet of copper to capture solar and cosmic radiations. The angled extension on the right side is a well to hold reagents, such as Biodynamic Preparations, for broadcasting.

The rendering on the left is of a design made by Ag Energizers, of Tiger, Georgia.

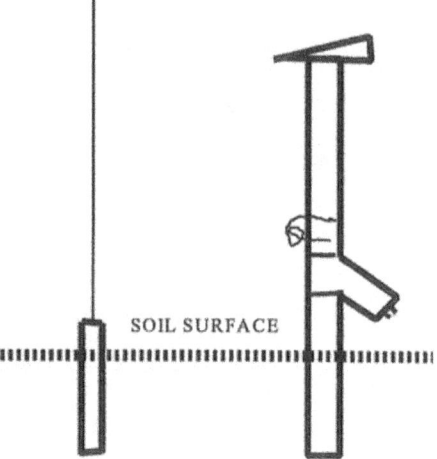

Graphic adapted from: Kuepper, George. 1998. Plants, Soils, Earth Energy, & Radionics. GAIA, Goshen, Arkansas. p. 152.

The Ag Energizer Pipe

In the photo, the author is holding an Ag Energizer field broadcaster made in the early 1990s.

Like the Hieronymus Cosmic Pipe, the body of the Ag Energizer unit is made with PVC pipe. The metal antenna is 8.5 feet in length. The body can accommodate several reagents or intent cards.

Though influenced by the Hieronymus tower, it is probably not correct to call this a Cosmic Pipe, since that name is specifically associated with the Hieronymus design.

Sometimes booster units are available to enhance the power of personal instruments. Such a booster was available for the Rogers unit. I was fortunate to acquire one several years ago. It will be shown in a photo later in this book. Unfortunately, the Rogers instrument and its booster are no longer being manufactured.

Some practitioners have found creative means for powering up their instruments. Marty Lucas has spoken of using mirrors placed beneath their rubbing plates. Others are trying magnets, pocket polarizers, crystals, and tensor rings. While I've not explored these "homemade" power boosters myself, they deserve mention. They might be the key to making some of the alternative psychotronics broadcasters more suitable for agricultural work.

The solution for those who plan frequent broadcasts to large land areas is usually to purchase more powerful radionics instruments. Alternatively, some choose field broadcasters.

Cosmic Pipes and Field Broadcasters

Sometime in the mid- to late 1980s, T. Galen Hieronymus and his team began marketing a psychotronics device called the *cosmic pipe*, designed to broadcast beneficial energies to agricultural landscapes, specifically to the soil.

Noted biodynamic farmer and researcher Hugh Lovel worked with a cosmic pipe on his farm for ten years. He upgraded the design to balance the atmosphere as well as the soil. This advancement blended well with the full range of traditional BD Preparations, which include both soil *and* atmospheric sprays.[20] Further details on his experiences, including construction plans are published in *Acres U.S.A.* magazine.[21]

---

20 Lovel, *Quantum Agriculture: Biodynamics and Beyond*, pp. 158–159.
21 Lovel, *Ten Years with a Cosmic Pipe*, pp. 12–15; Lovel, *A Cosmic Pipe Update…Farming the Atmosphere*, pp. 20–23; Lovel, *Stimulating Soil and Air*, pp. 20–25; Lovel, *Field Broadcasting 25 Years On*, pp. 58–62.

7

More about Witnesses and Protocols

In the chapter "Psychotronics," I discussed the concept and role of witnesses. In this chapter, I address the practical aspects of witnesses, including how to obtain and store them.

Traditional Witnesses

When working with humans or animals, traditional witnesses include blood, saliva, hair, and fur. Traditional witnesses for plants can be pieces of leaf, stem, or bark. If working with a large grouping of plants, as in a field or garden, a representational sampling sometimes works. Dowse to determine whether you've collected enough to make an effective witness.

Obtaining one of the best and easiest plant witnesses requires a bit of foresight. When you plant a crop, save a small sample of seeds from each packet, bag, or lot used. These seeds are entangled on the quantum level with seeds from the same sources and are excellent witnesses throughout the life cycle of the crop.

Traditional soil witnesses are a bit less complicated. Take representative sample *cores*[1] from the field or garden, much as you would for a conventional soil test, and combine them. Mix well and set aside a small amount as a witness. As with representational

---

1   *Cores* refers to soil samples 6 to 8 inches deep obtained with a soil probe. You can get equally suitable sample using a good shovel or spade.

sampling of plants, it is advisable to dowse to determine whether you've collected enough cores to make an effective witness.

When you collect such witnesses, avoid compromising them with the oil and sweat from your fingers. Use clean containers, rust-free tweezers, scissors, or knives and plastic or nitrile gloves.

I often store witnesses in glassine envelopes. This keeps them free of contamination, and they can be wiped free of stray fingerprints using a cloth or paper towel moistened with distilled water. Glassine is considered neutral so you can leave the witness in the envelope when placing it in an instrument or broadcasting device.

Like many practitioners, I also use lead-free glass vials, tubes, and beakers to hold and store witnesses. Like glassine, these can also be inserted into an instrument without compromising analyses or broadcasts. Glass is also easy to clean.

Photographic Witnesses

Instant photos are among the most common witnesses in current use. By "instant," I mean cameras and film that produce analog images within minutes after exposure. Chemical processing occurs within the film itself and there is no need for a laboratory.

In instant photos, the silver crystal emulsion in the film captures not only the physical image of the subject, but a subtle and holographic one as well, thus creating a very effective witness.

Despite that, I avoid using photos if possible. Instant film is expensive and, when working with a single plant or a small cluster of plants, it is not hard to obtain traditional witnesses. The situation is different when you need witnesses of entire gardens, large plantings, or crop fields. In those circumstances, instant photos are a godsend.

If you are fortunate, you can capture the image of an entire plot, garden, or field in a single photo. That is optimal but seldom easy to achieve without something like a camera-mounted drone. Usually, the lay and shape of the land or a poor vantage point prevents this. Sometimes I stand on a tractor, stool, or stepladder to capture as much of the land in the photo as possible. Such antics occasionally help, but only a little.

You frequently need to take multiple photos to capture large areas fully. The following graphic, borrowed from one of my earlier books, shows how you might take just two photos to cover a single large field. The images will overlap, but that is no concern.

Photo Witnesses of a Large Area

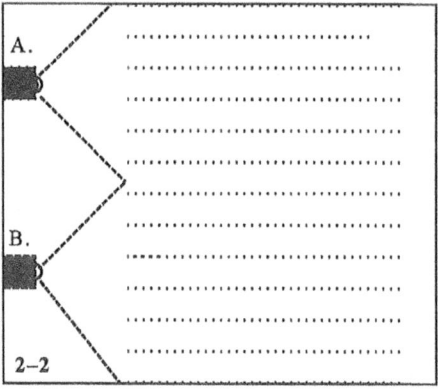

Image from Kuepper, George. 1996. Plants, Soils, Earth Energy, & Radionics. Goshen, Arkansas. p. 18.

If the fields are punctuated with small hills, swales, and depressions, you might want to take additional pictures, especially if large swaths are not visible in your overview photos. Since taking extra pictures can be expensive and time-consuming, consider this workaround. It entails using the sun's position to reinforce your intent process when you take a photo witness. In the following graphic, the photographer should be positioned so that the sunlight floods the landscape in front of the camera. By consciously taking the photo with the sun thus positioned, the witness will capture obscured and hidden field areas that are not visible in the image.

More about Witnesses and Protocols

Photo Witnesses of Expansive Area

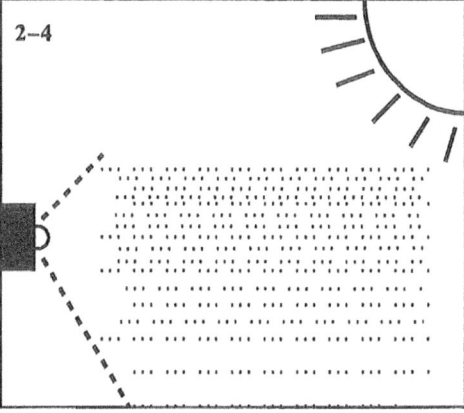

Image from Kuepper, George. 1996. Plants, Soils, Earth Energy, & Radionics . Goshen, Arkansas. p. 19.

We can use reverse logic when making a photo witness to isolate a block or a few rows of crop, or when your vantage point forces you to include a lot of unwanted background. Consciously placing the sun behind the camera, as shown in the next graphic, clarifies your intent to limit the area linked to the witness.

Photo Witnesses of Isolated Area

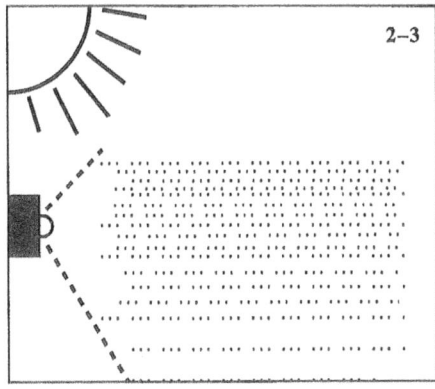

Image from Kuepper, George. 1996. Plants, Soils, Earth Energy, & Radionics . Goshen, Arkansas. p. 18.

By combining certain witnesses, you can also narrow your focus to target more effectively the specific plants or field area(s) you want to treat. For example, a photo witness of a field (with or without crop residues or cover crop) can be combined with a traditional witness of the seed to be planted there. I use such combinations routinely to plan for liming and fertilization. After seeding, the same witnesses will also serve for monitoring crop health and growth, for planning side-dress or foliar fertilization, and for psychotronics broadcasts to the growing crop.

For much of my career in radionics, Polaroid 600® was the instant film and camera of choice. It was so popular that several instruments were designed with slotted wells sized specifically to accommodate the photos. Unfortunately, in 2008, Polaroid ceased production of the 600 series and the psychotronics community began casting around for replacements.[2]

Around 2009, Kelly Research Technologies (KRT) compared several instant analog films and inkjet-printed photos as witness options. They concluded that all of the tested options worked to varying degrees, but Fuji Instax® film came closest to approximating the qualities of Polaroid 600®.[3]

I discovered, quite recently, that Polaroid has resumed production of its 600 film and cameras. At this time, they are not widely available. I have seen them only in specialized camera shops. It appears that the chemicals used in the film may have been changed, so it is unclear whether the new photos perform with the same integrity as the original.

Unless traditional witnesses are contaminated or compromised in some other way, they seem to sustain the link to their subject for a long time—often longer than practitioners will need them. Photo witnesses, on the other hand, are not so long-lived. Why this is so, I don't know. Perhaps it's because they are more easily contaminated or that they degrade more easily from the electromagnetic pollution that surrounds us. Perhaps their chemical constitution changes with time.

---

2   Anon., "Polaroid Film," *The Radionic Homestead Report*, vol. 21, no. 2. Mar./Apr. 2008, p. 2.

3   Anon., "Comparing Photographic Witnesses," *Kelly Research Report*, vol. 4, no. 2, summer 2009, pp. 7–9.

Surrogate Witnesses

A *surrogate* is, by definition, a substitute or a proxy. A *surrogate witness*, then, is a substitute for a traditional or photographic witness. Practitioners do not speak much about them anymore, but it was once common for some working with human subjects to write their names on slips of paper and use them for psychotronic analysis and broadcasting. Those pieces of paper with inscribed names are the classic examples of surrogate witnesses. Considering that the main purpose of a witness is to aid in establishing a resonant link with a subject, it is not too hard to understand that writing a person's name on paper would certainly bring him or her more clearly to mind, and thus achieve a resonant link.

Several of my radionics instructors discouraged us from using surrogate witnesses, especially when working with human subjects. Compared to standard witnesses, surrogates require more engagement and focus by the operator. As a result, they make the practitioner more vulnerable, especially if the subject has complicated issues of a psychic nature.[4] Consider this situation similar to the hazard of medical professionals catching a contagious disease from their patients. Many of us call this a form of *radionic backlash*—the unintended negative consequences we sometimes face when assisting others. It *is* something to take seriously.

The subjects we're addressing in this book, however, are plants, soils, or landscapes. Unless one is working with a graveyard or a site considered haunted, there is little to fear from using surrogates in the manner I'm suggesting.

Maps as Surrogate Witnesses

My wife and I live on four and a half acres of semirural land in northwest Arkansas. It contains our house, pets, several

---

4   Steve Westin, "Beginning class in Radionics," Mercedes, Texas, July 3–6, 1986.

outbuildings, a quarter-acre garden, manicured areas, and some wild landscape. I won't call it a farm but, to be consistent with my ongoing education in biodynamics, I view it like an *organism*. It has clear but permeable boundaries. Thought is given to what enters and leaves the property, especially when it comes to plant nutrients and the soil itself. I try to capture and utilize as much soil and plant nutrition as possible from the sun and the air through photosynthesis, nitrogen fixation, and deep-rooted plants. I concentrate much of that nutrition into the food garden. This means transferring a lot of fallen leaves, grass trimmings, and wood chips from elsewhere on the property.

I use two main hand-drawn maps as surrogate witnesses for my homestead and garden. The homestead map covers my entire property. It somewhat resembles the graphic that follows. (Because my penmanship is atrocious, the maps in this book are all computer-generated. Computer-drawn surrogates *can* work under certain circumstances, but I urge you to make hand-drawn maps, especially at the beginning.)

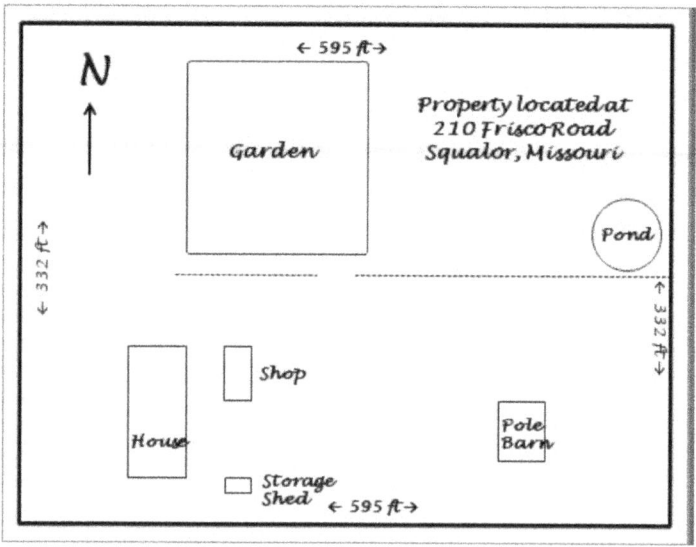

More about Witnesses and Protocols

As with intent cards, use clean, unmarked, white paper or card stock with India ink or a graphite pencil, always avoiding contamination of the witness. I suggest following up with lamination or sealing in a glassine envelope because surrogate map witnesses are ones you are likely to use frequently.

A homestead or farm map witness should have the following features:

- A clear indication of the property location. I show a physical mailing address, but you could also use a legal description or GPS coordinates. What is important here is that it brings focus to the correct location.
- Measurements of the boundaries. This clearly delineates the property and confines your analyses and broadcasts. Be as precise as possible.
- Approximate location and prominence of significant features, such as buildings, ponds, fields, and so on. Precision is *not* important here—approximate sizes and proximity are sufficient. Just be sure to include anything that significantly defines the property and aids focus.
- The surrogate should be small enough to fit the instrumentation you use for analysis and broadcasting.

I use my homestead map witness primarily to assess the need for BD Preparations and to broadcast them. I want their beneficial effects throughout my entire homestead. (These preparations will be discussed later.)

I made a separate surrogate map witness for my garden. Even though it appears on my homestead map and receives the benefit of broadcasts made to the whole property, there are additional things I want to do with the garden that require separate analyses and broadcasts.

The following graphic shows a garden similar to my own. The key features for this witness map are essentially the same as those

Psychotronics and a Biodynamic Garden

for the homestead surrogate, except that here the dimensions of the garden should be even more precise.

Surrogate Witness for Garden or Field

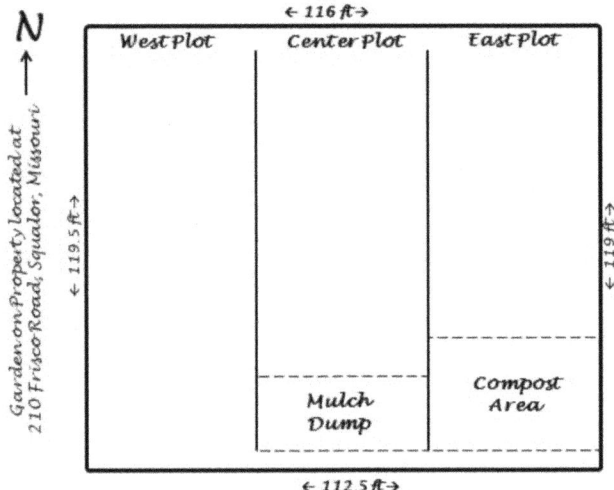

More about Surrogate Map Witnesses

Before beginning to draw a surrogate map witnesses, I encourage you to dowse whether it will work for you. Consider framing your question something like this: "Will a hand-drawn surrogate map of my property at [specify location] serve as an effective witness for psychotronics analysis and broadcasting?" Or similarly, "Will a hand-drawn surrogate map make a suitable psychotronics witness for my garden?" If the responses are NO, plan to use traditional or photographic witnesses.

I generally discourage using computer-printed witnesses unless one is closely and personally linked to the subject property. I currently use computer-drawn maps for my homestead and garden, but I've lived here for more than twenty-five years and bonded well with my property; I am responsible *for* it and responsible *to* it. I would never consider using a surrogate map witness for property belonging to anyone else. As for yourself, once you have confidence in your dowsing, you can make your own determination.

Alternative witnesses you might consider include aerial photos, such as those provided by the Farm Service Agency (FSA) or Google Earth. I've not used either, but there is much anecdotal evidence that FSA photos, at least, have worked well, especially for radionics pest control, and for managing Hieronymus pipes and Quantum Field Broadcasters. The wide availability of civilian, camera-mounted drones will almost certainly expand our options for photographic witnesses, though I've not yet tried those, either.

Witness integrity is integral and important to psychotronic work. Always dowse the witness suitability before proceeding. I dowse using the following language: "Is this witness suitable for analyzing and broadcasting to the subject?"

The Protocols of Psychotronics

When I speak of protocols, I'm referring to recognized procedures, rules, and accouterments considered necessary for psychotronics. Protocols include using witnesses, but apply to so much more.

When I began studying radionics, I was taught to do things in ways that reminded me of a microbiology lab—ways that ensured we would not compromise our work and experiments. Radionics uses laboratory protocols for the same reasons scientists and technicians use them: to prevent contamination and compromise. However, there is a downside to laboratory-like protocols. If we adopt their use literally on the same terms, they can mask the true nature of psychotronics and lead to confusion through obvious contradictions and inconsistencies.

To illustrate what I mean, I will use the example of the personal photo witness. When beginners take their first training in radionics, they commonly use themselves as practice subjects, and the instructor snaps their picture. Before taking the photo, the students are asked to remove jewelry, glasses, and pocket contents. They then stand against an unadorned wall while the photo is snapped. Based on our understanding of laboratory protocols, we do all this to exclude everything that might compromise a subject's photo witness. But if you give it some thought, you'll realize that the subject is still wearing clothes with buttons, zippers, sequins, and so on. Furthermore, the unadorned wall certainly contains dust, dirt, microbes and others' fingerprints. As an example of laboratory technique, it fails completely. However, as an example of witness collection for psychotronics work, it is excellent!

While psychotronics protocols mimic laboratory procedures, they are used for *ritual* purposes; ritual clarifies one's *intent* to isolate the energy field of the subject from all else around it. Rituals assist us in clarifying and supporting our intentions. This is why they work. As Donald Tyson, the author of *Ritual Magic* wrote, "Ritual is only an instrument to direct the mind."[5]

Therefore, the acts of removing jewelry from a subject and directing him or her to stand against a bare wall do not physically eliminate all contaminants. The ritual, however, makes the intention clear, and that is what counts in psychotronics.

---

5   Tyson, *Ritual Magic: What It Is and How to Do It*, p. 17.

8

Recycling Frequencies: An Application of Psychotronics

In *Report on Radionics*, Edward W. Russell describes the accomplishments of UKACO, a company that provided psychotronic pest control to crop farmers around the middle of the last century. One of its principals was Curtis P. Upton, whose explorations went beyond pest management. The following paragraph is especially intriguing:

> If he [Upton] merely wanted to stimulate the growth of a plot or a crop, he would put the leaf or photographic "key" (witness) on the collector plate and switch on the instrument for five or ten minutes. This was repeated about once a week, more often—even daily—in the growing season. In some mysterious way, the "radiation pattern" of the leaf, or "key," would be strengthened and transmitted back to the plant or the crop. This invigorated the tree or crop: the green color would become darker and new growth longer than in the case of untreated plants.[1]

Later writers have called this technique *recycling frequencies.*

A Modified Approach

Russell's description of Upton's procedure is clear and self-explanatory. Around 2015, I made modifications to it that are still under evaluation, but which I'm comfortable sharing. In addition to a basic radionics instrument, it requires a homeopathic potentizer.

---

1 Russell, *Report on Radionics: Science of the Future*, p. 52.

Most Hieronymus-style instruments either have built in potentizers, or make them available as an add-on.

My experimental procedure follows:

- Obtain a suitable plant or crop witness(es).
- Neutralize the instrument and potentizer wells.
- Place the witness(es) in the input well and confirm their integrity.
- Add a small amount of distilled water or a few neutral sugar pellets to a clean lead-free beaker (150 ml beakers easily fit Kelly and Rogers instruments; neutral sugar pellets are available from homeopathic suppliers).
- Place the beaker in the potentizer well. Toggle the instrument rate bank "on"; set the rate dials to 00–100; set the potency dial to zero. If the potentizer unit has a *phase* or *polarity* switch, toggle it "in."
- Toggle the broadcast switch "on" to imprint a *base-level energetic copy* of the witness onto the water or sugar pellets; wait three minutes; toggle "off."

Radionic Potentization: Base Level

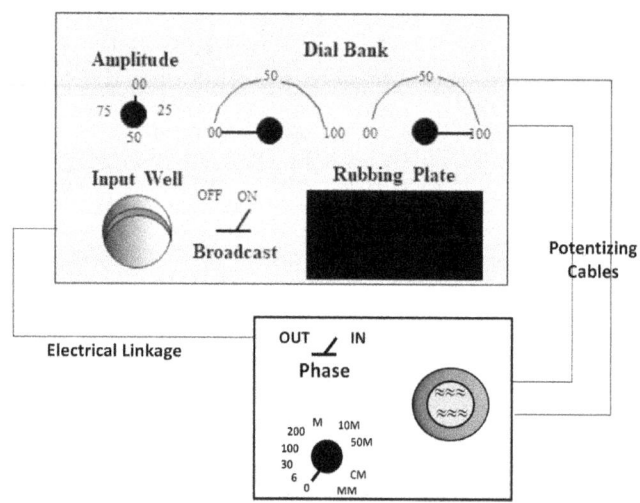

- Next, stroke or hold a pendulum over the rubbing plate while rotating the potentizer dial to seek "the *first* homeopathic level at which this subject should be potentized." A firm stick or change in pendulum action indicates the correct level. Leaving the potentizer on this setting, once again toggle the broadcast switch "on," wait three minutes, and toggle "off."

Seeking Homoeopathic Levels

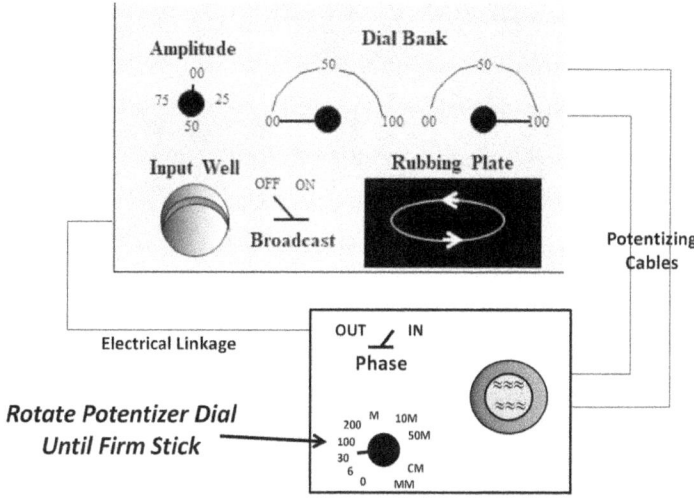

- Repeat the process, seeking "the *second* homeopathic level at which this subject should be potentized." When found, potentize at the new level as done before. (You should continue looking, but I seldom find more than one or two homeopathic levels.)
- Once completed, switch the beaker and its potentized contents from the potentizer well to the input well, placing it alongside the subject's witness. If the potentizer is not integrated with the main instrument, unplug it and set it aside.
- Ensure that the rate bank remains set to 00–100. Dowse to see if it is safe and advisable to broadcast.

- Toggle the broadcast switch "on" and dowse for the broadcast time required.

Recycling Frequencies Broadcast

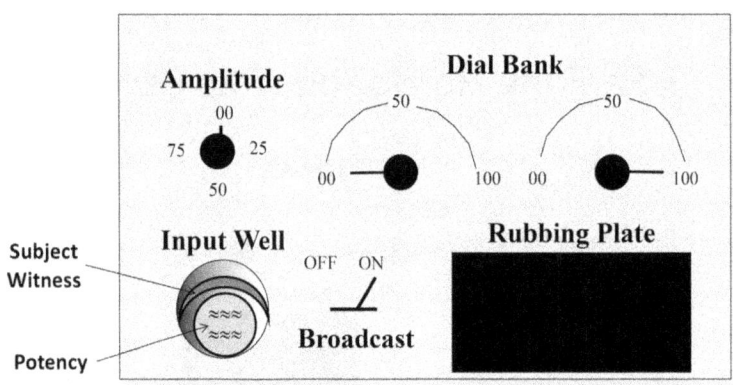

Should you try to recycle frequencies, remember that *my* modifications are not fully evaluated and might be flawed. Dowse to determine the best way to proceed.

A Segue to the Next Chapter

I am introducing *recycling frequencies* at this point because it is a "plant-positive" in contrast to "pest-negative." It is about enhancing plant vitality, *not* about killing pests or disease organisms. It is important to make this distinction. One of the recurring themes of this book is the growing recognition that the majority of plant-disease infections and pest infestations occur when crops, soils, and agricultural ecosystems are out of balance. Therefore, our most reliable long-term solutions is to restore a healthy balance. This is the subject of the next chapter.

9

Rationale for the Plant-positive Approach
In the Beginning, There Was Psychotronic Pest Control

In the previous chapter, I referred to UKACO, a company that had success controlling field crop pests during the 1940s and 1950s. For many who practice agricultural psychotronics, it remains a historical touchstone. I cover the saga of UKACO in detail in the appendix, so I won't do so here, except for a few points that are pertinent to this chapter:

- UKACO used psychotronic instrumentation and protocols to kill and/or repel insect pests from large fields of agronomic crops.
- The operators are said to have "painted" some form of pesticide onto an aerial map witness featuring the targeted fields.
- Documentation is anecdotal but convincing. It seems that UKACO *did* achieve success controlling several insect pests in different crop fields.

The story is compelling! It goes far in validating the belief that psychotronics really does things—things that people can see and measure. But there is an added appeal—intended or not—that comes from the science fiction imagery the tale evokes. Imagine voracious pests "zapped" from a distance, using mysterious *black box* technology.[1] This is seductive stuff!

---

1   Many people used the term *black box* to describe radionics instruments. I've not discovered a definitive reason for this, though numerous opinions abound.

Psychotronics and a Biodynamic Garden

Setting my snarky sci-fi comments aside, there is no question that psychotronically killing or repelling pests is far better for the environment and human health than blanketing crops with poisons. Anyone researching or applying such methods should be applauded and carry on with their work. I have used some of these methods myself, with at least modest-to-good success. Furthermore, I expect I'll use them again! In this book, however, I'm looking at plant pest and disease management from another vantage. I'm adopting a notion that is a cornerstone of organic agriculture: that truly healthy plants and agroecosystems naturally resist and suppress pests and diseases.[2]

Organic and Biodynamic Philosophy

While I had read of this idea in my earliest readings on organic farming, I did not encounter a solid demonstration until my last year working with Washington University. While I'd been on many healthy organic operations, all with few pest and disease problems, it wasn't until 1978 that the message really hit home, when I visited Ralph and Rita Engelken on their organic farm in northeast Iowa. The Englekens believed strongly in that old organic adage. They were strong advocates of crop rotation,[3] and grew a mix of field crops on a planned six-year sequence (see graphic). However, Ralph and Rita also grew a separate thirty-acre test plot. This plot had been planted with corn for the previous thirteen years—an example of continuous cropping common through much of the region. However, unlike their conventional neighbors, the Engelkens used no insecticides, and instead of commercial fertilizer, they made annual applications of three to five tons of biodynamic compost.

---

2 Hainsworth, *Agriculture: The Only Right Approach*, pp. 191–215.

3 Crop rotation is the sequencing of different crop on the same land year to year. Farmers rotate crops to aid in soil health and fertility management, to manage diseases and pest insects, and to reduce erosion.

Rationale for the Plant-positive Approach

Engelken's Six-Year Crop Rotation

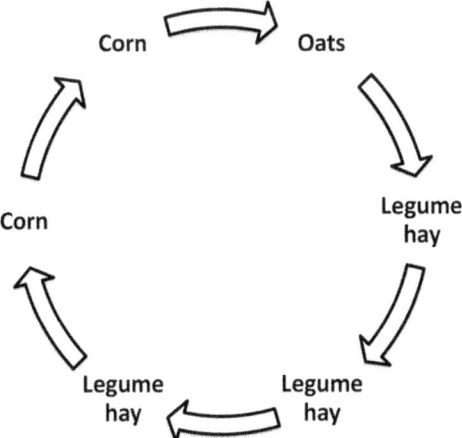

Adapted from: Engelken, Ralph & Rita. 1981. The Art of Natural Farming and Gardening. Barrington Hall Press, Greeley, Iowa. p. 135.

The test plot produced good yields, but this was not the most significant observation; the corn did not lodge[4] or show other ravaging by northern and western corn rootworms. In current times—when corn varieties are genetically engineered to resist rootworms—this would not seem significant, but in the late 1970s it was!

Adult rootworm beetles lay eggs in cornfields every year. Their larvae emerge and feed on the plant roots. During a second year of corn, the larvae do observable damage, but it is limited because their populations are too low. By the third year of continuous growing, however, their numbers are high and roots get badly pruned. This causes corn plants to topple even in moderate winds.[5] I had seen a few examples of this around the corn belt; yields are reduced

---

4   The term *lodge* describes breaking or bending plant stems, which directly reduces crop productivity and makes harvesting difficult or even impossible. Lodging commonly results from winds and is made much worse when roots or stems suffer pest insect and disease damage.

5   Engelken and Engelken, *The Art of Natural Farming and Gardening*, pp. 111–112.

and machine-harvesting becomes very difficult. At the time, conventional growers made annual applications of insecticides like carbofuran (Furadan®) and terbufos (Counter®), to control the larvae; organic growers, by contrast, rotated their crops, never growing corn more than two years in a row.

As for the Engelken's thirteen-year continuous corn demonstration, the means by which rootworm damage was averted is hard to explain other than through soil–crop–agroecosystem health and vitality.[6] The key to achieving this seemed to be the routine use of biodynamic compost, along with other good organic practices.

Later, I had another reminder of the traditional organic claim linking disease and pest resistance to plant vitality. It was part and parcel of my efforts to manage the Kerr Center blueberry planting using psychotronics. My earliest radionics analyses left me no doubt that nutritional issues and soil imbalances were at the root of the problems, especially the high levels of insect pests and plant diseases. Using radionics, I formulated foliar fertilizers made from traditional organic materials like fish and seaweed products, natural mineral suspensions, humic extracts, vitamins, and biologicals. However, I also included technical grade soluble fertilizers, food-grade phosphoric acid, chelated micronutrients, and surfactants. Insect pests, mites, and plant diseases vanished promptly, demonstrating that they were only symptoms. The decline in the planting was reversed almost immediately.

Obviously, my inclusion of several synthetic materials in my sprays precluded calling my methods or crops "organic" in a strict sense, but I had made a point. When pests and diseases took their leave, it was in direct response to my correction of nutritional problems in the plants. So, although the methodology and inputs did not meet organic certification standards, I had demonstrated a connection between health and vitality, along with the ability of plants to naturally resist insect pests and diseases.

---

6   Ibid., p. 112.

Nature's Garbage Collectors

The belief in a linkage between plant vitality and resistance to pests and diseases is not solely the purview of traditional organic growers and a few radionics practitioners. There are two overlapping schools of thought and research on this theory: *predisposition*[7] and *trophobiosis*.[8] Both disciplines note that when plants are stressed they stop "building" proteins and begin "reducing" them. This floods plant tissues with free nitrogen compounds and amino acids. These simpler materials are the preferred foods of herbaceous insect pests, like aphids, which lack the enzymes to break down whole proteins.[9,10] This being true, healthy plants deny insect pests the digestible food they need—food they could easily find on sick or overly stressed plants.

This has led to an interesting take on the function of pest insects in the environment. A few growers call them nature's "sanitation workers" or "garbage collectors," since they are charged with removing vegetation that is damaged, genetically flawed, or otherwise unable to cope with environmental stresses.

The stresses that affect plant physiology and bring on diseases and insect pests are various. In the case of the Kerr Center blueberries, I was able to identify the causal stress as nutritional. There were imbalances of some crop nutrients and deficiencies of others. However, it is important to recognize that there were causes behind

---

7 Coleman and Ridgeway, "Role of Stress Tolerance in Integrated Pest Management," p. 126, in Knorr, *Sustainable Food Systems*.

8 "The theory of trophobiosis has to do with how plant nutrition affects plant health and what makes a plant susceptible or resistant to diseases and pest attacks (Chaboussou, *Healthy Crops: A New Agricultural Revolution*, p. 2.

9 Eliot Coleman, quoted by Kim Stoner and Tracy LaProvidenza, "A History of the Idea that Healthy Plants Are Resistant to Pests," in Stoner, *Alternatives to Insecticides for Managing Vegetable Insects: Proceedings of a Farmer/Scientist Conference* (NRAES–138), Dec. 6–7, 1998, p. 4.

10 Anon., "Pests Starve on Healthy Plants." *Ecology Action Newsletter*, Willits, CA. May 1999, pp. 3–4.

the nutritional causes. These included inadequate and ill-advised fertilization, poor pH management, and the consequences of too many synthetic herbicides, fungicides, and insecticides. All of these stresses are *agricologenic,* meaning they are caused by farming methods—the things that growers do.

We have choices: we can attack the symptoms with more pesticides and commercial fertilizer, often making the problems worse, or we can look for and correct root causes. Traditional organic and biodynamic growers lean toward the latter; they evaluate their growing systems and methods to find what they might be doing wrong. Some remedial action, like natural pesticides, could be needed to get by, but reaching the long-term objective of health and vitality often requires profound, well-considered changes that do not create and leave new problems in their wake.

And Psychotronics?

Addressing the cause behind symptoms is in the nature of radionics as it is applied to humans. It is, however, also true for agriculture. T. Galen Hieronymus, in the *"credimus"*[11] (introduction) to *Cosmiculture,* outlines his philosophy and approach to using radionics–psychotronics for farming and gardening. He writes:

> Disease, unwanted insects, undesirable plants are simply indications of conditions, in that environment, conducive to their existence at a particular time and place. Change those conditions by enhancing the environment for the desirable, and the reason for the undesirable ceases to exist. Reagents are incorporated within the Cosmiculture system that will, at once, enhance the vitality of the desirable *and* reduce the vitality of the undesirable.[12]

---

11 Roughly translated from the Latin, *credimus* means "what we believe."
12 Hieronymus, *Cosmiculture,* p. 1.

Rationale for the Plant-positive Approach

This statement of belief mirrors the ideas common to predisposition, trophobiosis, and the traditions of organic agriculture and biodynamics.

10

Biodynamics

I first heard of biodynamics (BD) in the late 1970s while working for Washington University and studying organic farming. It intrigued me, but it was much too complicated for me to wrap my head around. Over the years, I continued to read the odd article or two. I'd meet someone knowledgeable on occasion who helped explain some things. I even attempted to use some of the BD Preparations on a few occasions, but didn't understand what I was doing or why.

I *did* have success working with the natural rhythms of the sun, moon, planets, and constellations. When managing Kerr Center's blueberry farm, I used almanacs to schedule foliar fertilization. I believe, strongly, that it added to our success. Later, when experimenting with cover crops on the Cannon Project, I also used almanacs to schedule mowing operations. I chose trimming dates that increased growth when I wanted the vegetation to recover quickly, and trimming dates that retarded growth when I wanted to kill and plow the cover crop down. Again, I felt that this helped. In both instances, I relied, primarily, on Llewellyn's annual *Moon Sign Book*.[1]

It wasn't until a few years ago, however, that I committed myself to the serious practice of biodynamics. My years of working with psychotronics set the stage quite well. Both psychotronics and BD have similar energetic worldviews.

---

1   Llewellyn, Woodbury, Minnesota (www.llewellyn.com).

As you read further, it will be clear that I have much less experience with biodynamics than I have with psychotronics. You might note that some of my observations and conclusions lack seasoned insight and sophistication. Well, "it is what it is." Everyone starts somewhere. Understand that what I am presenting here is a progress report on my efforts to fuse psychotronics and biodynamics. If I waited on my mastery of biodynamics, this book would never have been written.

What Is Biodynamics?

Biodynamics is a spiritually based approach to farming and gardening. It grew out of Anthroposophy—a philosophical movement established by the Austrian philosopher, social reformer, and clairvoyant Rudolf Steiner (1861–1925).[2] It constitutes a spiritual path of knowledge rooted in the European idealism and built around the teachings of Aristotle, Plato, Thomas Aquinas, and others.[3] Anthroposophy is often referred to as "spiritual science."[4] It assumes "an objective, intellectually comprehensible spiritual world, accessible to human experience."[5]

Steiner laid out the basis for biodynamics through a series of eight lectures delivered in 1924 in Koberwitz, Silesia (now part of Poland).[6] Several landowners and farmers had asked him for insights into how they might farm in a more sustainable and healthy way.

---

2   Biodynamics is sometimes called an *initiative* of Anthroposophy. In truth, it was Steiner's last significant initiative before he passed away in March 1925 (McKanan, *Eco-alchemy*, p. 11).

3   Robert Mays and Sune Nordwall, "What is Anthroposophy? Waldorf Answers on the Philosophy and Practices of Waldorf Education" (waldorfanswers.org/Anthroposophy.htm).

4   Hans van Florenstein Mulder, "A New Approach to Scientific Investigation," p. 145, in *Biodynamics: New Directions for Farming and Gardening in New Zealand* (New Zealand Biodynamic Farming and Gardening Association).

5   See https://en.wikipedia.org/wiki/Anthroposophy.

6   Steiner, *Agriculture*.

Steiner's audience was already tuned to Anthroposophy, so he spoke freely about esoteric matters that the uninitiated would not easily understand, either then or now.

The lectures became the basis for a new approach to agriculture—one that embraced science, but within a wider, spiritual context. And what has evolved—biodynamics—is not a stereotyped system, but a unified approach to agriculture that relates the ecology of the earth as an organism to that of the entire cosmos.

That biodynamics is, first and foremost, "spiritually based," should not and cannot be discounted. If we do so, we lose sight of the rationale for its esoteric practices *and* its embrace of practical organic culture. The failure to recognize the spiritual and philosophical origins can lead us to falsely conclude that biodynamics is "merely organic agriculture with a peculiar twist."[7]

An Emphasis on Quality

The landowners and farmers who requested guidance from Rudolf Steiner were concerned about declines in soil, seed, food, and feed *quality*. While they were certainly interested in good yields, that did not appear to be their prime concern. This is a crucial point. Steiner did not promise that alchemical preparations and esoteric practices would produce higher yields. That was never the main objective.[8] Better health and higher quality were!

Quality can mean many things. It can manifest as improvements in crop health, as evidenced by natural disease suppression, fewer insect infestations, superior grain test weights, better keeping quality, and reduced nitrites in produce.[9] These are metrics often used by biodynamic researchers.

---

[7] Hugh Courtney, "So that the Earth May Be Healed: An Introduction to Biodynamic Agriculture," p. 10, in Steiner, *What Is Biodynamics?*

[8] In practical application today though, the preparations sometimes do increase yields (Walter Goldstein, "Experimental Proof for the Effects of Biodynamic Preparations," *Biodynamics*, Sept./Oct, 2000, pp. 6–7).

[9] Ibid., pp. 7–8.

In modern agriculture and food science, researchers commonly measure nutrient levels, nutrient proportions, protein content, amino acid balance, fiber content, contaminants, and toxins. From this information they infer food and feed quality. Such research has been done on organic and biodynamic produce, though not nearly enough. Perhaps most notable is the work of clinical nutritionist and naturopath Virginia Worthington. She is best known for her meta-analysis of collected research, comparing the nutritional qualities of organically and conventionally grown foods.[10] Her findings validated many organic and biodynamic claims, demonstrating overall higher levels of Vitamin C, magnesium, and phosphorus in organic produce.

A longtime member of the Biodynamic Association, Worthington published an article in the *Biodynamics Journal,* asserting that the quality of BD produce was, at the very least, consistent with that from other organic systems, and that biodynamic produce appears to have a slight edge in Vitamin-C levels.[11] She confesses, however, that the pool of research results was, at that time, too small to make further nutrient claims.

In her article, Worthington also references fourteen livestock feeding trials in which organic and conventional regimes were tested. In eight of them, animals provided with organic feed performed better, exhibiting less illness, recovering faster when ill, showing better fertility and sperm motility, laying more eggs, suffering fewer stillbirths, and exhibiting better survival rates among live births. In *none* of the studies did conventional feeding outperform organic. All of the trials in which biodynamic feed was used were among those in which organic performance was significantly better.[12]

---

10   Virginia Worthington, "Nutritional Quality of Organic Versus Conventional Fruits, Vegetables, and Grains," *The Journal of Alternative and Complementary Medicine*, vol. 7, no. 2, 2001. pp. 161–173.

11   Virginia Worthington, "Nutrition and Biodynamics: Evidence for the Nutritional Superiority of Organic Crops. "*Biodynamics*. July/Aug. 1999, pp. 22–25.

12   Ibid., pp. 22–25.

Such feeding trials provide better indicators of food and feed quality than reductionist laboratory findings. Observing and measuring animal performance when they are fed biodynamic feedstuffs is much more objective and definitive. One can readily see whether the feeding regime works better or not; no inference is required. In such trials we look at the combined results of better vitamin and mineral content, higher protein quality, purity, and many other factors. To me, these mensurables *and* nonmeasurables constitute what we call *life force*.

I found the following explanation of life force on the website for the Biodynamic Association, UK, and consider it one of the better definitions:

> The biodynamic understanding of vitality is "life force"—that universal energy that's a true reflection of positive health, be that of soils, plants, animals or people. As the poet Khalil Gibran said in his poem, The Prophet, it's "life longing for itself." Chinese Medicine, for example, expresses it as "chi," Ayurvedic medicine as "prana," but the concept is universal. It embraces physical, mental and spiritual health (and) is the foundation of everything that biodynamic farmers and gardeners strive to achieve.[13]

In traditional radionics we routinely use an analysis rate (09–49) to measure what we call General Vitality (GV). It represents the overall condition of the entire plant, animal, or human system, including the physical, the mental, and the spiritual aspects.[14] We use readings of General Vitality as benchmarks, against which we compare the measurements of individual parts, functions, and attributes of a

---

13 Jessica Standing, "Is Vitality Merely Feeling Great and Full of Life?" The Biodynamic Association UK, 2017 (https://www.biodynamic.org.uk/vitality-merely-feeling-great-full-life).

14 Anon., "General Vitality (9–49): What Does it Really Mean," *Radionicsspectrocom*. Mar. 8, 2017 (https://radionicsspectro.com/2017/03/08/general-vitality-9-49-what-does-it-really-mean).

system. If we discover that one or more of these is significantly out of balance, we take steps to bring them back in line.

A measurement of General Vitality is something more than just a baseline. Radionics practitioners regularly observe that the GV reading of a subject typically rises as imbalances are corrected. This tells us that higher General Vitality readings usually indicate greater coherence, whereas lower ones point to more dissonance.[15] On the practical level of horticulture, low-vitality, "dissonant" plants and agroecosystems do not perform well. They are more easily thrown for a loop when challenged by inclement weather, harmful radiations, and pollutants.

Biodynamics and Organic Growing Methods

Biodynamic growers typically use practices that are common to other organic gardening and farming systems. These methods originated with traditional farming systems from around the globe, but primarily from an early-twentieth-century approach called *humus farming*.[16] Humus farming held to the belief that healthy food and feed could only come from healthy soil, and that healthy soil depended on the vitality and diversity of organisms living in it. These included bacteria, fungi, nematodes, protozoa, earthworms, insects, macro- and micro-flora and fauna, and plant roots.

We now use the term *soil food web* to describe the totality of these organisms—their abundance, their diversity, and their relationships. The "web" consumes and processes wastes, fixes nitrogen, suppresses disease, creates humus, and ensures nutrition to all its living entities, including crop plants. According to humus farming, we should model our fields and gardens on natural

---

15 Marty Lucas, "Analysis of Radionics," Feb. 5, 2019 (https://radionics.us/analysis-of-radionics).

16 George Kuepper, 2010, "A Brief Overview of the History and Philosophy of Organic Agriculture," Kerr Center for Sustainable Agriculture, Poteau, Oklahoma, pp. 2–3 (http://kerrcenter.com/wp-content/uploads/2014/08/organic-philosophy-report.pdf).

ecosystems to conserve and enhance the soil food web. We need to return organic matter—manure, compost, natural fertilizers, plant and food wastes—to the soil; this is the nourishment that drives it.[17]

Humus farming avoids inputs and substances that suppress or kill soil organisms. *This* is why the early organic movement prohibited most pesticides and salt-based fertilizers; they did damage to the soil food web. It was not until the 1960s that concerns about pesticide residues on food, and the wider environmental impacts of agricultural chemicals, became the compelling rationale for organic methods.[18]

As we've learned more about the soil food web, we have learned that the traditional organic practices of tillage and cultivation can be highly counterproductive. Inverting and disturbing the soil kills many soil organisms. It destroys earthworm tunnels and fungal networks, compacts the soil, reduces humus levels, and leaves land open to erosion. It seems that much of the benefit from manuring, green-manuring, and residue management is spent compensating for the destructive effects of tillage and crop cultivation.

As we've come to grips with this reality, organic and biodynamic growers are reconsidering their tillage practices. Many are shifting to reduced cultivation, mulching, and non-chemical no-till systems. Others however, still adhere to the tillage-based systems they find reliable, and work to make them more sustainable through other means. I feel that *biointensive mini-farming* as taught by John Jeavons is an excellent example of the latter. It was first introduced to the United States by Alan Chadwick (1909–1980) in the 1960s as the *French intensive-biodynamic method*. Biointensive mini-farming balances deep tillage (double-digging)

---

17 Ibid.
18 Ibid., pp. 3, 9.

with compost, made from soil, weeds, residues, and compost cover crops grown in a planned rotation with vegetables and grains.[19]

Shortlist of Organic Cultural Practices

Composting	Organic no-till options
Applying manure	Pheromone applications
Mulching	Using natural fertilizers
Tillage	Trapping
Cultivation	Hand-picking pests
Hand-weeding and hoeing	Fire and flaming
Screening and row covers	Rotating crops
Scouting for pests	Cover cropping
Using natural pesticides	Green manuring
Choosing resistant plant varieties	Beneficial habitats
Timed planting	Introducing natural predators

Cosmic and Earthly Influences

The science of ecology helps us understand our world in terms of relationships. We've learned that the smallest of things—the gain or loss of a single plant, animal, or microbial species, for example—can lead to profound changes throughout an ecosystem, a hemisphere, and even the whole planet.

Biodynamics expands our understanding of ecology to include the influences of the moon, planets, and constellations. However, the science and traditions surrounding it are complex, and I readily get out of my depth. Despite that, I'll share a few insights that I've found helpful:

- In biodynamics, celestial observations are based on their relationship to planet Earth—a *geocentric* perspective. This is opposed to the *heliocentric* view which places the sun in its known place at the center of the solar system. Biodynamics

---

19 Anon., "History of Ecology Action," Ecology Action/Grow Biointensive (http://www.growbiointensive.org/about_history.html).

does not deny astronomical observations and facts, but it does recognize that plants and other life forms experience the moon, planets, and constellations from the perspective of *this* planet.[20]

- Most of us are aware of moon phases—full, new, waxing, waning. We associate these with varying lunar effects on plants. There are more, including ascending and descending moon, as well as the moon's distance from Earth (e.g., *apogee* and *perigee*).[21]
- Biodynamics also considers the effects of the zodiac—the ring of twelve star constellations visible in the night sky. There are slight differences between the signs of the zodiac common to astrology and the astronomical observations made by BD researchers and practitioners.[22] This is one reason why biodynamic calendars and almanacs can have recommendations that differ from gardening guides based on astrology.
- Cosmic influences can significantly affect the effectiveness of BD Preparations, and some practitioners schedule applications accordingly.[23]

One of the characteristics of biodynamics I love and respect is the willingness to adopt promising new technologies and ideas, while holding tightly to tried and true practices of enduring value. Continuing to acknowledge and work with earthly and cosmic cycles is an excellent example of this. Biodynamics refuses to throw *that* baby out with the bathwater.

---

20 Gita Henderson, "The Workings of the Moon, Stars and Planets," 1989, p. 141, in New Zealand Biodynamic Farming and Gardening Association. *Biodynamics: New Directions for Farming and Gardening in New Zealand.*

21 Ibid., pp. 131–134.

22 Ibid., pp. 134–137.

23 Ibid., pp. 139–140.

Treating the Farm as an Organism

While biodynamics recognizes oneness with the cosmos, it also has a strong consciousness of boundaries. Biodynamic practitioners treat their farms or homesteads as single-cell organisms contained within semipermeable membranes. They recognize flows of nutrients and energy within the organism, as well as ongoing exchanges with the outside environment. This encourages greater self-reliance and self-sufficiency by reducing off-farm inputs and minimizing waste and pollution. It also encourages the development of community. This is one reason why so many BD growers choose community supported agriculture (CSA) marketing.

The Biodynamic Preparations

Hugh Lovel—a BD practitioner of many years—wrote a brief but well-crafted definition for the BD Preparations, to wit: "Natural medicines used on the land to impart the forces necessary for a balanced and healthy environment."[24]

There are nine original preparations, numbered 500 to 508, which emerged from Steiner's agriculture lectures. Growers have been using them for many decades; they are probably *the* defining practice of biodynamics.

The preparations resemble homeopathic medicines. They are used in very small amounts. About one and a half ounces of 500 Horn Manure, when properly stirred into four gallons of water, is sufficient to treat one acre; one-twentieth of an ounce of 501 Horn Silica, similarly stirred into four gallons of water, also treats one acre. Because the amount of substances used is so small, one cannot ascribe efficacy to nutrient content. The preparations are *not* fertilizers and should not be viewed in those terms. Though it is only an analogy, I find it helpful to think of them as *catalysts*.

---

24 Lovel, *A Biodynamic Farm*, p. 5.

Psychotronics and a Biodynamic Garden

In physical chemistry, a catalyst is a substance that accelerates a chemical reaction without itself undergoing any permanent chemical change.[25] *Enzymes* are good examples of biological catalysts. They facilitate and speed up natural processes. Note that the various preparations shown in the following table "work with," "attract," and "stabilize" nutrients; there is *no* claim that they "provide" or "deliver" them.

But unlike enzymes, which function biochemically on the dense physical level, BD Preparations work on the etheric plane, with what Steiner called *etheric formative forces*. The preparations "create conditions under which plant and soil become sufficiently sensitive to react to and absorb the incoming stream of life from the cosmos."[26]

The preparations also have much in common with homeopathic medicines in that both are *potencies*—physical substances that are imprinted with *energy information*. However, they differ in the ways they are prepared. The processes for making the preparations (see the table on page 117) are quite arcane. Suffice it to say that these alchemical procedures have much to do with enlisting terrestrial and cosmic forces to aid in developing healthy soil and plants.[27]

Even if one is intrigued by the processes, making BD Preparations is not very easy. It requires assembling a number of plant and animal materials that are hard for many of us to come by. Therefore, many BD farmers and gardeners purchase preparations from individuals and organizations—for example, the Josephine Porter Institute[28]—which specialize in making them.

---

25 See https://www.dictionary.com/browse/catalyst?s=t.

26 John Soper, "Studying the Agriculture Course," cited in Storl, *Culture and Horticulture*, 1976, p. 345.

27 Koepf, *What is Bio-Dynamic Agriculture*. pp. 14–16.

28 Josephine Porter Institute for Applied Biodynamics, 652 Thompson Rd. SE, Floyd, VA 24091 (www.jpibiodynamics.org).

Biodynamics

Nine Original Biodynamic Preparations

500 Horn Manure	Spray made from manure fermented in a cow horn. Buried in soil from late fall to early spring.	Promotes root activity, soil microbes, N-fixation, seed germination
501 Horn Silica	Spray made from silica packed into a cow horn and buried in the soil from late spring to early fall	Stimulates photosynthesis; enhances color, aroma, flavor, keeping qualities
502 Yarrow Blossoms (*Achilea millefolium*)	Used in compost; flower heads from yarrow are packed into a male stag or deer bladder to ferment; suspended first in sun from early summer to mid–late fall; then buried until middle of next summer.	Works with sulfur and potassium; attracts trace elements
503 Chamomile (*Matricaria recucitata*)	Used in compost; flowers from German chamomile are stuffed into bovine intestines and buried from mid–late fall to early spring	Works with sulfur and calcium; stabilizes nitrogen
504 Stinging Nettle (*Urtica dioica*)	Used in compost; stinging nettle plants are buried, surrounded by peat or well-rotted sawdust from mid-summer until late summer of the next year	Works with hemoglobin and chlorophyll; enlivens the soil
505 Oak Bark (*Quercus robur*)	Used in compost; ground oak bark is stuffed into the cranial cavity of a domestic farm animal and immersed from fall through early spring in a location with trickling water	Works with calcium and nitrate; helps prevent plant disease
506 Dandelion Flowers (*Taraxacum officinale*)	Used in compost; dandelion flowers are packed into a bovine peritoneum and buried from mid-fall to early spring	Stimulates relationship between silica and potassium; fruiting
507 Valerian Flowers (*Valeriana officinalis*)	Used in compost; fermented juice made from an extract of fresh valerian flowers	Works with phosphorus and blossoming
508 Horsetail (*Equisetum arvense*)	Spray made by boiling common horsetail	Works with silica and ripening; antifungal

Table adapted from Walter Goldstein, "Experimental Proof for the Effects of Biodynamic Preparations," *Biodynamics*, Sept./Oct. 2000, p. 6.

Psychotronics and a Biodynamic Garden

Several of the original preparations require *dynamization* before applying, most notably 500 and 501—*Horn Manure* and *Horn Silica*, respectively. Dynamization is a prescribed stirring procedure that transfers the imprinted energy information from the preparation to water, so that it can be sprayed. We might see dynamization as *secondary potentization*; the first occurs when energy information is first imprinted onto the preparation substance—manure, quartz, or herb; the second occurs when it is transferred from the substances to water.

The stirring procedure begins promptly after placing the preparation in a bucket or barrel of water. We stir the mixture in one direction, rapidly enough to create a vortex; then, suddenly reverse direction to collapse the vortex and create a new one that flows in the opposite direction. Alternate stirring is continued for an hour to ensure complete dynamization.[29] When done, the mixture should be strained to remove particulate matter and prevent clogging spray nozzles.

It is beyond the scope of this book to dwell further on these details. Besides, they are well covered in the biodynamic literature. However, here are a few particulars worth mentioning. With regard to spray preparations:[30]

- Use unpolluted, lukewarm water for dynamizing. Rainwater is optimal. Tap water must be cleansed of chlorine as thoroughly as possible.

- Stirring vessels should, if possible, be well-cleaned wooden barrels or glazed earthenware crocks. Some proponents suggest metallic containers as the next-best option. However, author and longtime practitioner Beth Weiting disagrees. She

---

29 Lovel, *A Biodynamic Farm*, p. 13.
30 Most of the details provided here were adapted from Koepf, *Koepf's Practical Biodynamics*, pp. 74–76.

Biodynamics

recommends using old plastic containers if wooden barrels or buckets are not available.[31]

- Sprayers should be dedicated for biodynamic use, if possible. Purchasing a new sprayer is a fine idea, but it can be expensive. Thoroughly clean used equipment but *never* attempt to use sprayers that were once used for synthetic herbicides—especially chlorophenoxy materials like 2,4-D or 2,4,5-T. It is nearly impossible to remove all the vestiges of these poisons and the smallest amounts remain toxic to plants.
- As an alternative to mechanical sprayers, one can also apply the spray preparations using a clean bucket and an inexpensive whisk broom or brush. Walk the field or garden carrying a bucket of the dynamized water. Dip the brush into the water and, with a flick of the wrist, distribute it in as wide a pattern as you can. Ideally, each square foot of ground will receive at least one droplet. This is another practice that makes little sense to conventional thinking. However, if you grew up in the Roman Catholic Church, as I did, you will recall that priests often bless the congregation by flicking holy water over everyone, using an *aspergillum*.[32] All present receive the blessing through the intent of the ritual. If one understands that holy water and liturgical blessings also carry energy information, it should not be hard to understand how a dynamized preparation can be effectively applied by a similar unorthodox means.

With regard to the compost preparations:[33]

---

31 Wieting, *Nature Spirits*, p. 35.

32 An aspergillum is a handheld whisk or other liturgical tool designed to hold a small quantity of holy water that is released as a pattern of droplets with the flick of a wrist.

33 Most of the details provided here were adapted from Tompkins and Bird, *Secrets of the Soil*, pp. 386–387.

- Biodynamic composting generally follows the guidelines for standard aerobic pile and windrow composting.
- Unlike compost starters, which are usually distributed or layered throughout a pile as it is built, the traditional BD preps are inserted according to a prescribed pattern after the pile is constructed. Use a broom handle or pry bar to make channels into the pile or windrow, and then insert a pinch or hand-rolled pellet of the preparation. The exception is 507, the valerian preparation, which must first be dynamized. Half of it is poured into the designated channel; the remainder is sprinkled over the top of the pile or windrow.

Inserting Traditional BD Compost Preparations in a Windrow or Pile

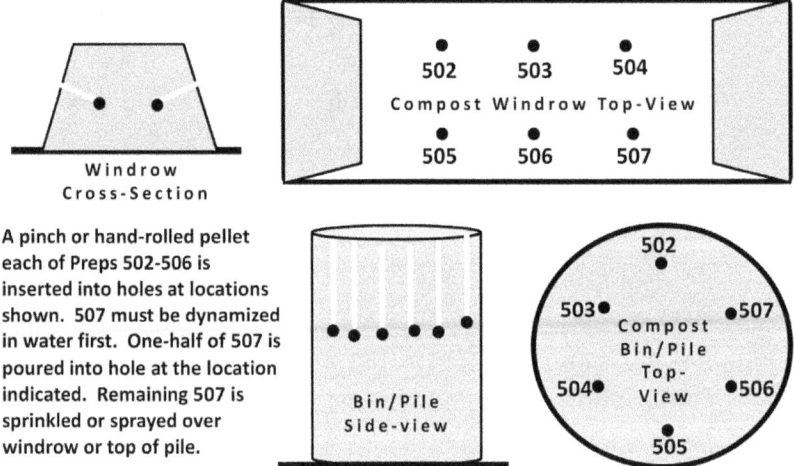

A pinch or hand-rolled pellet each of Preps 502-506 is inserted into holes at locations shown. 507 must be dynamized in water first. One-half of 507 is poured into hole at the location indicated. Remaining 507 is sprinkled or sprayed over windrow or top of pile.

Asking the Right Questions?

I have heard several people ask why biodynamics, a decidedly *organic* growing system, would burden itself with esoteric practices and alchemical substances that make the scientific community and the general public question its credibility. It is fair to ask that. And if you do, I hope this book will give you some answers.

But to be intellectually honest, one should also ask why biodynamics, a spiritually conscious approach to farming and nature, *chooses* organic principles and methods over modern chemical and genetically engineered inputs. Is organic more life-affirming?...more respectful of the planet?...more conscious of the rights of nature and future generations of humanity? Again, I hope you find some answers here.

Nature Spirits, Devas, and Etheric Formative Forces

In the paragraphs to follow, I will discuss the most enigmatic and bewildering topic in practical biodynamics—devas, elementals, and what Rudolf Steiner called *etheric formative forces*. At the time I began using psychotronics at Kerr Center's blueberry farm, I did not understand such things, nor did I see any relevance in them. It was some time later, after reading Machaelle Small Wright's books, that I became cognizant of their reality beyond myth and fairy tales. And now, as I delve ever deeper into biodynamics, the importance of the intelligent beings of nature has become real and critical to what I am doing. In fact, I consider these subjects the most important ones I cover in this book.

Though I'd hoped to avoid a lot of metaphysical and esoteric language and details, it is difficult to do so when writing about nature beings. Perhaps this is because I am still new to biodynamics and fail to grasp so many of the concepts.[34] But that won't keep me from forging ahead. I'll begin with *elementals*—the beings that some call *nature spirits*.

---

34   For anyone wishing to look deeper, the text I find most helpful and frequently cite is Tompkins, *The Secret Life of Nature: Living in Harmony with the Hidden World of Nature Spirits from Fairies to Quarks.*

Psychotronics and a Biodynamic Garden

There is a lot of folklore surrounding nature spirits, as well as a growing body of contemporary writing. However, it is the work of theosophists, anthroposophists, and alchemists that provide us with our best insights and understanding. As I do with so many difficult subjects, I rely heavily on interpretive literature. It is mostly from those writers and teachers that I've acquired my conception of elemental beings and what they do.

My earlier impressions of elementals were that they simply "hung about" in natural settings, the way teenagers hang out at malls and convenience stores. They are there, but we don't know much about what they're doing. I know that sounds sarcastic, but I believe a lot of people entertain similar images.

Thankfully, my thinking on this has matured. Elementals have function and they have responsibilities. They are the spiritual beings that bridge the etheric and dense physical levels. With guidance from yet higher spiritual–angelic beings, elementals marshal etheric substance into the templates on which living physical matter is built. *They* are responsible for the formation and maintenance of living matter.

Rudolf Steiner frequently wrote and spoke about *etheric formative forces*. If what I've written so far is valid, then we might logically assume that elementals either direct or otherwise work with these etheric formative forces—perhaps ordering or pushing them around in sensible ways. But this is only one interpretation. Another argues that elementals *are* the etheric formative forces. Furthermore, Steiner chose the less-controversial term and neutral image of "forces," because this would be more palatable to scientists and the public than the traditional visions that elementals or nature spirits evoke.[35] R. Ogilvie Crombie (1899–1975), also known as "Roc," was one of the key personalities associated with the Findhorn community. He wrote, "We can, if you like, regard those elementals

---

35 Ibid., p.120; also, Ernst Hagemann, "The Elementals as Helpers in Farm and Garden," pp. 31-32, in Erbe, *Hugo Erbe's New Bio-dynamic Preparations*.

as forces, energies, because that is what they are. So, just as the particles forming the atom may not be matter at all, but consist of whirls of energy, these beings are really energy, an intelligent energy, which can become personalized."[36]

Crombie wrote further that pure elementals are "light" bodies—vortices of energy that come from higher angelic realms.[37] He is telling us that elementals are not physical at all, but higher beings that take on etheric form to do what they do in creating living matter.

Alchemists taught that the elementals are comprised of the basic ethers: *warmth, light, chemical* (or tone), and *life*. Each of these ethers corresponds to the traditional alchemical elements of *fire, air, water,* and *earth*, respectively.[38] It is important that we not confuse or equate the four alchemical elements with the physical chemical elements of the periodic table. The alchemical ethers and elements describe the nature and mechanics of the etheric and, possibly, higher planes of reality—*not* the physical plane. The alchemical elements are *not* physical substances but *etheric forces* that maintain matter in its fiery, gaseous, liquid, and solid states.[39]

About Theosophy

Theosophy is most commonly associated with the movement founded by Helena Blavatsky (1831–1891) and Henry Steel Olcott (1832–1907) in New York in 1875 as the Theosophical Society.[40] It teaches that knowledge of God can be achieved through spiritual ecstasy, direct intuition, and special individual relations.

Theosophy claimed to teach the essential truths found in all religions so that it might appeal to Christians, Buddhists, Hindus,

---

36 Crombie, *Encounters with Nature Spirits*, p. 162.
37 Ibid., p. 87.
38 Tompkins, *The Secret Life of Nature*, p. 112.
39 Ibid.
40 See https://www.lexico.com/en/definition/theosophy.

Parsis, Hebrews, and Muslims, who, in fact, thronged to the society while retaining their own religion.[41]

Modern Theosophy may be defined as a synthesis of the essential truths of religion, science, and philosophy.[42]

While it emerged in the West, Theosophy is heavily influenced by Eastern religion and mysticism. It has sought to be a philosophical bridge between East and West. Rudolf Steiner was intimately involved in Theosophy before branching off and establishing Anthroposophy, which offers a more Western- and Christian-influenced spiritual path.

About Findhorn

Findhorn is a spiritual center and community in northeast Scotland, started in 1962, by Peter (1917–1994), Eileen Caddy (1917–2006), and Dorothy Maclean (1920–2020). All three had dedicated themselves to following a disciplined spiritual path. The community gathered great attention through its prolific garden and its work with elemental and devic beings. Today, the Findhorn Foundation is an NGO associated with the United Nations Department of Public Information. It is a holder of UN Habitat Best Practice designation and is co-founder of the Global Ecovillage Network and Holistic Centres Network.[43]

Each nature spirit is forged from one of the four ethers and is thus bound to the associated alchemical element. Fire elementals—called salamanders—are composed of warmth ether and work with the element of fire. Steiner accorded this element a special status, considering it fundamental to all things physical and spiritual. In plants, fire is associated with procreation and seed development.[44]

Sylphs and fairies are the elementals composed of light ether that work with the air element. Nymphs, dryads, naiads, nixies, undines, sprites, and watermen are names used to describe nature

---

41 Tompkins, *The Secret Life of Nature*, p. 31.
42 Cooper, *Theosophy Simplified*, p. 13.
43 Findhorn Foundation website: https://www.findhorn.org/about-us/.
44 Tompkins, *The Secret Life of Nature*, pp. 56, 118–120.

spirits associated with water. Those associated with the earth element are recognized by many names and forms including fauns, gnomes, dwarfs, brownies, knockers, and kobolds.[45, 46] The nature spirit Pan is a powerful faun and is treated as a minor "god" within the elemental kingdom.[47] Pan is the *highest representation of the earth element beings.*[48]

Elementals shape the natural world, but they are not self-guided and do not act in isolation; they take direction from astral entities called *devas*. In the *vibrational hierarchy* of spiritual beings, devas are considered angelic and often described as the lowest order of angels. Devas are responsible for the blueprints of creation—the templates that nature spirits use when making living matter. This hierarchical arrangement exists because devas have too high a vibration to work directly on the etheric level and must enlist the services of elementals to fashion the physical world.[49] So, like architects, devas provide the blueprints and guidance to the elemental craftsmen, who build accordingly.

In this regard, William Bloom, of the Findhorn community, wrote:

> Devas have an exact sense of what the perfect plant should be and that the changes wrought by the interference of weather, of other plants, of soil conditions, of animals, and of people are all inputs to which the deva fluidly adjusts, always clearly holding the sense of the perfectly fulfilled plant. Thus, no matter what external interference there might be, the plant

---

45 Pogačnik, *Nature Spirits and Elemental Beings: Working with the Intelligence in Nature*, pp. 119–158.

46 Baan, *Lord of the Elements: Interweaving Christianity and Nature*, pp. 139–141.

47 Crombie, *Encounters with Nature Spirits*, pp. 11–17.

48 Pogačnik, *Nature Spirits and Elemental Beings*, p. 129.

49 Tompkins, *The Secret Life of Nature*, pp. 124, 126.

itself is still enveloped by the ideational matrix, held by the deva, into which it may grow.[50]

In addition to constructing the etheric bodies of plants, elementals infuse them with energy received from devas that they transmute to plant-usable forms.[51] As Peter Tompkins (1919–2007), coauthor of *The Secret Life of Plants*, wrote: "In the occult view of life on earth, elemental spirits lie hidden behind all that constitutes the physical, sense-perceptible world, making it, with their effort, truly alive."[52]

So, What about DNA?

DNA (deoxyribonucleic acid) is the self-replicating material within the chromosomes of nearly all living organisms. It carries genetic information.[53] DNA was first isolated in 1869 by Friedrich Miescher (1844–1895). Its molecular structure was described in 1953 by Francis Crick (1916–2004) and James Watson (b. 1928).[54] Since conventional science accepts that DNA contains all of the information needed to build and maintain living organisms, we might wonder whether devas and elementals are relevant and play any role at all.

What we need to remember is that DNA and chromosomes are aspects of our physical plane, whereas elementals and devas exist and operate essentially on the etheric and astral levels. We *might* reason that biochemical DNA derives from the activities of nature spirits and devas and is guided by them, and that it is a fundamental aspect of life forms on the physical plane that nature beings ultimately bring about.[55] Viewing DNA in this manner does not deny, minimize, or contradict the demonstrated facts of biochemical or physical science; it merely provides a bridge for understanding its spiritual origins and relationships.

---

50   Ibid., p. 127–128.
51   Crombie, *Encounters with Nature Spirits*, pp. 60–61.
52   Tompkins, *The Secret Life of Nature*, p. 112.
53   See https://www.lexico.com/en/definition/dna.
54   See https://en.wikipedia.org/wiki/DNA.
55   Crombie, *Encounters with Nature Spirits*, pp. 147–148.

Several sources make it clear that nature beings have a different path of spiritual evolution from human beings.[56] This is consistent with what I learned as a young Roman Catholic. We did not speak about human beings becoming angels. Rather, if people reached a high level spiritually, we called them *saints*. Humans and angels clearly had their own, very different, roles within creation.

Folklore is replete with tales of people seeing and speaking with gnomes, fairies, and other nature spirits. Such interactions seem rare these days, and we might wonder why? To begin with, Western culture discounts such folk stories—dismissing them as imaginary or outright fiction, and insisting that these events never actually occurred. This is further supported by the assertion that, before humankind fell under the sway of modern materialism, people welcomed spiritual phenomena into their lives. But today, for those of us who believe nature beings exist and would love to catch a glimpse, we might wonder why we still do not see them. I must admit that I have not yet had that pleasure.

The principal reason we don't see these omnipresent beings is that even the densest etheric bodies are too rarified to reflect any of the rays within the spectrum of ordinary light.[57] Apparently, some level of *clairvoyance* is needed.[58]

Theosophic literature describes several levels of clairvoyant sight. The most basic is *etheric sight*. It allows one to see lower-order elementals as well as the health aura surrounding the physical human body. *Astral sight* is clairvoyance on a more refined level. It allows vision of still higher vibrational forms.[59] Some people are born with clairvoyant capabilities—perhaps more frequently than

---

56 Tompkins, *The Secret Life of Nature*, pp. 61, 142; Moore, *Divining Earth Spirit: An Exploration of Global and Australasian Geomancy*, p. 123; Crombie, *Encounters with Nature Spirits*, p. 86.

57 Tompkins, *The Secret Life of Nature*, p. 34

58 *Clairvoyance* is the ability to discern objects or perceive matters beyond the range of ordinary perception (*Merriam-Webster*).

59 Tompkins, *The Secret Life of Nature*, pp. 32–34.

Psychotronics and a Biodynamic Garden

we realize. Sadly, as children, we are urged to ignore and abandon these gifts because they don't square well with materialism or the dogma of Western Christianity.

It is possible to develop clairvoyant skills. Steiner discusses this in *How to Know Higher Worlds*, originally published in 1904. More recently, a book based on Steiner's teachings has been published: Shelli Joye's *Developing Supersensible Perception*. But even for those who do not study or prepare themselves for supersensible perception, glimpses of nature beings or other spiritual phenomena may still occur. As with many ESP events, sometimes they just seem to happen. It is widely held that seeing elementals and devas is not something that human beings control. Rather, glimpses must be permitted by these spiritual beings themselves.[60] It is believed that the main reason sightings are so rarely granted is because most elementals avoid humanity due to our widespread denial of their existence and our destructive behavior toward the environment and the planet.[61]

Another mystery surrounding elementals is why, when they *do* allow humans to see them, do they look the way we expect them to—like the Green Man, Tinker Bell or the Seven Dwarfs? Elementals do not have either fixed physical *or* etheric bodies. When visible, they adopt etheric forms based on human *thought-forms* and expectations.[62] Based on art, myths, legends, and popular culture, our conscious and subconscious states already have notions of what elementals *should* look like. You might say that *we see what we expect to see* if we are fortunate enough to "see" at all.

In *Culture and Horticulture*, anthropologist Wolf Storl writes that using imagery from ancient traditions to describe otherwise unexplainable phenomena serves a number of beneficial purposes. One of these is understanding that what we perceive is more than

---

60 Crombie, *Encounters with Nature Spirits*, pp. 79–81.
61 Tompkins, *The Secret Life of Nature*, pp. 60–61.
62 Ibid., p. 43; Moore, *Divining Earth Spirit*, p. 130.

the sum of its deceptively mechanistic parts.[63] Rather than forcing ourselves into the paradigm of reductionist math, chemistry, and classical physics, we view the world around as enchanted and alive. Those of us who view the world in this way feel compelled to live more lightly on the land and treat the planet as something worth conserving, understanding, and respecting.

Traditionally, Western Christianity has had concerns about elementals, suspecting they might be evil. Should they be condemned along with anyone who consorts with them? Christianity has a long and complicated history with elementals. In an effort to dispel paganism during the Middle Ages, the image of Pan as part human and part goat became the Church's image of Satan. Other elementals have regularly been portrayed as fiends, imps, goblins, and devils. There is no doubt that the Church wanted to portray elementals as evil. It is our understanding, however, that elementals occupy the lower rungs of the angelic hierarchy and, as spiritual beings, *"must love."* They lack the free will to hate or do evil. Humans, on the other hand, have free will and can choose the dark side if so inclined.[64]

In *Encounters with Nature Spirits*, R. Ogilvie Crombie recounted a personal interaction with Pan on the small island of Iona in the Hebrides: "He [Pan] was smiling and said: 'I am the servant of Almighty God, and I and my subjects are willing to come to the aid of mankind, in spite of the way it has treated us and abused nature, if it affirms belief in us and asks for our help.'"[65]

This is not to argue that harmful beings don't exist and don't cause problems. They do and they can. In his book, Crombie relates an exchange he had with Pan regarding pseudo-elementals, also known as black goblins, imps, and the like. Pan denies that these entities are his "subjects." He asserts that they exist on a low astral

---

63 Storl, *Culture and Horticulture*, pp. 64–67.
64 Tompkins, *The Secret Life of Nature*, p. 108.
65 Crombie, *Encounters with Nature Spirits*, p. 20.

plane and have their own "god"—one that resembles Pan but is the true *nymph-chasing satyr,* which was the real model for the Christian Satan.[66]

In a chapter titled "The Question of Evil," Crombie also writes the following:

> I said I believed the elementals were part of the angelic hierarchy. Some people have often said, and this is something I have to touch on, that there are no bad nature spirits; there are no evil beings. But of course there are. I do not class them as true elementals, but they exist. There are several kinds. Again we have the possibility of humanity building up thought forms. We may do it by simply thinking evil thoughts. We could also do it, I believe, by writing. We could create such things as vampires: thought forms which could be activated by some sort of negative energy, and then you would get black goblins. I know this because I have seen them. I have on several occasions had to deal with them.[67]

About Thought Forms

With regard to thought forms, the following quote from the theosophists Annie Besant (1847–1933) and C. W. Leadbeater (1854–1934) is pertinent:

> Each definite thought produces a double effect—a radiating vibration and a floating form. "The radiating vibration" conveys the character of the thought, but not its subject. For example, serenity or devotion radiating from one person can stimulate similar vibrations in a nearby person who is receptive. The floating form is a strong and definite thought that has attracted energies from the mental and astral planes, and has become, for a time, a kind of independent living being.[68]

This quotation mirrors my understanding of *tulpas* and *tulpoidal phenomena*. A *tulpa* is a concept in mysticism that describes a being or

---

66 Ibid., pp. 86–89.
67 Ibid., p. 160.
68 "Thought-forms" (https://theosophy.wiki/en/Thought-Forms).

object created through spiritual or mental powers.[69] In *Mysticism and the New Physics*, author Michael Talbot (1953–1992) describes the experiences of Alexandra David-Néel (1868–1969) in Tibet. Using focused thought forms, David-Néel manifested a *tulpa* in the form of a phantom monk that appeared regularly during her overland journey. The *tulpa* was regularly visible to her, and sometimes to her fellow travelers as well. When she ceased her interest in the *tulpa*, she was surprised at how long the phantom continued to manifest.[70]

A Further Note on Biodynamics

We are living through a time when reality and truth seem "up for grabs." Far too many politicians, church leaders, celebrities, and others promote lies, hatred, biased narratives, and unfounded conspiracies that fly in the face of current, historical, and scientific realities. I *like* to believe I'm among those who seek clear-headed, sensible, and sometimes brutally honest views, but there are those who would debate that.

As I've spent time learning about the biodynamic movement and the people that comprise it, I find it more and more attractive. It is reassuring and delightfully ironic that a community that embraces notions of elemental beings, etheric forces, and celestial influences simultaneously acknowledges the glaring realities of climate change, systemic racism, and economic injustice in thoughtful, clearheaded, and practical ways. Conventional "wisdom" suggests that such people would be too airheaded to be rational about anything, but that is far from true.

As evidence of this, let me highlight an issue of *Applied Biodynamics*,[71] published by the Josephine Porter Institute. Three-fourths of that issue addresses climate change with sound science

---

69   *Tulpa;* see https://en.wikipedia.org/wiki/Tulpa.
70   Talbot, *Mysticism and the New Physics,* pp. 104–105.
71   Josephine Porter Institute, *Applied Biodynamics,* no. 93, spring/summer 2018.

and practical adaptations by biodynamic farmers. There are no denials; no excuses.

As further evidence, I point to the 2018 "Diversity, Equity, and Inclusion in the Biodynamic Association" statement of the Biodynamic Association and the BD community of North America.[72] It specifically acknowledges the stewardship of American land by indigenous peoples long before ethnic cleansing and colonization, and also the injustices of black slavery, white supremacy, and capitalist greed that drove an exploitative agricultural economy. No denials; no excuses.

I find solid kinship with people and organizations willing to confront both history and current events with great honesty and fairness. For this reason and in pursuit of spiritually based and biologically sustainable farming and gardening, I am pleased to count myself as part of the biodynamic community.

---

72 *Biodynamics*, no. 293, fall 2018, pp. 20–21.

11

Nature Beings, Psychotronics, and Co-creative Gardening

In this chapter, I detail how I interact and work with nature beings when using psychotronics. I have already mentioned Machaelle Small Wright, who has been a great influence on my efforts, and I will continue to reference her frequently. I encourage anyone who wishes to delve further into this subject to seek out her writings.[1]

Co-creativity

For Machaelle Small Wright, *co-creativity* involves human beings partnering with nature's intelligences (elementals and devas) to make something happen. "Something" can be just about anything. For example, I partnered with nature beings to create this book. The process is amazing! Our theme and objective—the "something" we address here—is a successful garden or farm.

When addressing groups of farmers and gardeners, I've often pointed out that what we create is not natural. Left to its own devices, nature will grow things other than those we choose. We can employ organic principles and methods to mimic nature, but we are still asserting our agenda. In this regard, Machaelle writes:

> Humans tend to look at gardens as an expression of nature. Nature looks at gardens as an expression of humans. They (gardens) are initiated, defined and maintained by humans.

---

1   See https://www.perelandra-ltd.com.

When humans dominate all aspects and elements of the life of the garden, we consider this environment to be human dominant. We consider an environment to be "nature friendly" when humans understand that the elements used to create gardens are form, and operate best under the laws of nature and when humans have the best intentions of trying to cooperate with what they understand these laws to be. When humans understand that nature is a full partner in the design and operation of that environment—and act on this knowledge—we consider the environment to be actively moving toward a balance between involution and evolution. (Involution being nature's dynamic): Providing balanced order, organization and life vitality for moving soil-oriented consciousness into form. (Evolution being the human dynamic): Providing definition, direction and purpose. In short, when a garden operates in a balance between involution and evolution, it is in step with the overall operating dynamics of the whole.[2]

Seeking a Balance between Involution and Evolution in a Garden

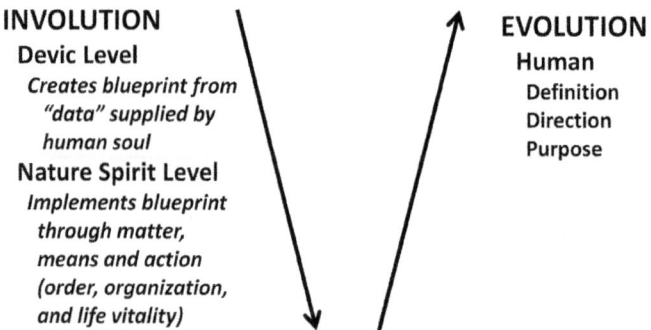

INVOLUTION
Devic Level
Creates blueprint from "data" supplied by human soul
Nature Spirit Level
Implements blueprint through matter, means and action (order, organization, and life vitality)

EVOLUTION
Human
Definition
Direction
Purpose

Adapted from: Wright, Machaelle. 1997. Co-Creative Science. Perelandra, Ltd., Jeffersonton, Virginia. p. 18.

Small Wright urges us to approach nature's intelligences in an ordered manner; she advises that we consciously acknowledge, and

---

2   Small Wright, *The Perelandra Garden Workbook* (ebook), Perelandra, Jeffersonton, Virginia, 2012, p. 6.

interact with, three kinds of beings—*devas* (nature's architects), *nature spirits* (the craftsmen and builders), and *Pan* (who serves as a bridge between devic and elemental realms).[3] She avoids specific references to elves, fairies, gnomes, and so on, preferring to see the whole of nature's intelligences as "a massive and flowing intelligence force."[4] I am following this advice and will refer to Machaelle's "intelligence force" using the collective terms *co-creative partners*, *co-creative beings*, or *nature beings*.

Psychotronic Connection and Communication with Co-creative Beings

In bringing psychotronics to this matter, I'm seeking a comfortable means for communicating with my co-creative partners to draw on their wisdom and guidance. I begin by formulating an intent statement similar to that shown in the next graphic. The phrasing is almost identical to that I use for my homestead. The key difference is that I specify the Shoreline Way Project in my homestead statement. Should you borrow the phrasing shown below, you might want to do something similar.

A General Intent Statement for Working With Co-Creative Partners

Engage Devas, Nature Spirits, and Pan for my garden and homestead landscape activities.

To utilize the intent statement, convert it into an intent card, using the protocols described earlier. If you can, I encourage creating a radionics rate and writing it on the card (see next graphic).

---

3   Small Wright, *The Perelandra Garden Workbook*, p. 13.
4   Ibid., p. 12.

Since I use such intentions often, I make several copies and laminate all of them.

Intent Card for Working With Co-Creative Partners

> Engage Devas, Nature Spirits, and Pan for my garden and homestead landscape activities.
>
> XX.XX—XX.XX

Scanned Radionic Rate For the Written Intention

Machaelle Small Wright's PKTT Method

In the Perelandra literature, Small Wright teaches a different means for communicating with co-creative partners. It is an adaptation of *applied kinesiology* or *muscle testing* that she calls *Perelandra Kinesiology Testing Technique,* or PKTT.[5]

PKTT tests muscle strength or weakness, which is another form of dowsing response, equivalent to finger-sticks or pendulum swings. Small Wright details this methodology in *The Perelandra Garden Workbook*.[6] PKTT is certainly as credible and reliable as any of the dowsing procedures I've described thus far. Some readers might find Machaelle's method more effective; if so, by all means use it!

---

5 See https://www.perelandra-ltd.com/PKTT-Perelandra-Kinesiology-Testing-Technique-C795.aspx.

6 Small Wright, *The Perelandra Garden Workbook*, pp. 47–56.

12

Balancing the Whole Farm and Homestead

To organize my approach to biodynamics and psychotronics, I've subdivided my efforts to target three nested levels:

1. The whole homestead or farm level
2. The garden or field level
3. The individual crop or plant level

These are not hard-and-fast divisions. The needs and activities on each of these levels vary, though in many cases they do overlap.

The Whole Homestead

My homestead includes the garden, flower beds, lawns, and wooded land. It includes land that I manage intensively and land that I've left to nature. It is at this all-inclusive level that I make the first and most frequent use of the BD Preparations, which I believe benefit all of the land in my care.

Using the Biodynamic Preparations

In the chapter on biodynamics, I described the nine basic preparations and their uses. Here, I am expanding the topic and discussing how I bring psychotronics into the picture.

Since most of my procedures from here on entail co-creativity, I need to describe how I arrange things psychotronically. When using a radionics instrument or a template, I place my witness(es) for the farm or homestead in the input well, alongside the intent card that I described in the previous chapter—the one that engages my co-creative partners. From this point on, I simply refer to this as the *Engage Card*. When using a pendulum without an instrument or template, I suspend it above the witnesses and *Engage Card* as I dowse. When stick dowsing, I like to place the witnesses and the Engage Card in front of me on the table or desk and rub the unoccupied smooth surface on the right (being right-handed). In all instances, I place my list of preparations conveniently to my left so I can finger-scan it with my left hand. I am now prepared to dowse to know what needed at the present time.

We need not dowse blindly when scheduling preparation applications. The BD literature gives us excellent guidance. In the northern hemisphere, the following is recommended:

- BD 500 Horn Manure—apply to the soil fairly early in the spring up until planting.
- BD 501 Horn Silica—apply to foliage in late spring and early summer, preferably in the morning.
- BD 508 Equisetum tea—apply to foliage frequently and as needed.[1]
- Compost preparations BD 502–507—use whenever making compost. However, they are sometimes individually indicated, and can be applied to soil, seed, or growing plants as a potentized spray, drench, or seed soak.

If you use the preparations according to this standard guidance alone, you should expect to be successful. However, by using psychotronics along with co-creative assistance, you can time your applications more effectively.

Here is an example: Assume it is early spring and the usual time to spray Horn Manure (BD 500). The grower can dowse to

---

1 Koepf, *Biodynamic Sprays*, p. 11–13; see also *Koepf's Practical Biodynamics: Soil, Compost, Sprays and Food Quality.*

Balancing the Whole Farm and Homestead

determine the optimal month, week, or day to make the application. If practical, he or she might dowse the optimal time of day for spraying as well. Since there can be benefits from spraying BD 500 more than once in a year, one can also dowse if and when additional applications might be done.

While I wrote earlier that I wanted all of the land and plants on my homestead to benefit from the BD Preparations, it is not always practical to physically spray preparations over everything. You should do only what is reasonable and concentrate on fields, garden, pastures, orchards, etc. It is only when one uses psychotronic broadcasting that reaching every square inch of your landscape becomes reasonable. I will discuss that later in this chapter.

More Preparations

Rudolf Steiner's lectures are the original source for the traditional preparations 500 to 508. However, as biodynamic farming and gardening has continued to evolve, growers and researchers developed some new formulations; they also created some completely new preparations. I am including a few of them here:

BD 501(F). Traditional BD 501 is made using ground quartz.[2] BD 501(F), by contrast, is feldspar-based. It is the recommended alternative to use on sandy soils.[3] My soils are not sandy, yet I regularly dowse 501(F) as the preferred form for my crops.

Pfeiffer's Biodynamic Field Spray. Originated in the 1940s by Dr. Ehrenfried Pfeiffer (1899–1961), this spray contains the BD Preparations 500, 502, 503, 504, 505, 506, and 507.[4] Rather than

---

2   Tompkins and Bird, *Secrets of the Soil*, p. 396.
3   "BD 501(F) Horn Silica Feldspar-Based (BD spray preparation)," Josephine Porter Institute: https://jpibiodynamics.org/collections/biodynamic-preparations/products/bd-501f-horn-silica-feldspar-based-bd-spray-preparation.
4   "Pfeiffer™ Field and Garden Spray," Josephine Porter Institute: https://jpibiodynamics.org/collections/biodynamic-preparations/products/pfeiffer

producing them by stirring, they need to be created through a twelve- to sixteen-hour procedure of moisture-activation.[5] In a 2017 webinar sponsored by the US Biodynamic Association, farmer and educator Stewart Lundy suggested that the Pfeiffer Field Spray is comparable to #500P, or *prepared Horn Manure*, which is exceptionally good for increasing soil humus levels.[6]

Biodynamic Compound Spray Preparation. This preparation has several variations and names, such as *barrel compost* and *cow pat pit*. The formulation I use comes from the Josephine Porter Institute (JPI) and is based on a recipe developed by famed biodynamic researcher Maria Thun (1922–2012). It contains the BD compost preparations 502 to 507, cow manure, basalt, and eggshells. It is particularly useful when one has limited access to BD compost. However, the compound spray preparation is not recommended as a long-term substitute for BD compost.[7] JPI recommends spring application ahead of BD 500; after that, it can be used as needed and is especially helpful when plants are stressed.[8] I typically lack enough compost and find this preparation exceptionally helpful.

Horn Clay. In the second of Steiner's lectures for the Agriculture Course, he spoke of clay as a mediator between "lime and silica," or between BD 500 and 501. He did not, however, go on to describe an additional preparation based on clay. Some argue that it was unnecessary since clay is used to plug the cow horns when preparing

---

%E2%84%A2-field-and-garden-spray-1.

5   "Pfeiffer™ BD Field & Garden Spray," Josephine Porter Institute, Floyd, Virginia, p. 2.

6   Stuart Lundy, "Climate Change, Carbon, and Biodynamics" (webinar), East Troy, WI: Biodynamics Association, Oct. 12, 2017.

7   "BC—Biodynamic Compound Spray Preparation" (adapted from Maria Thun's barrel compost recipe), Josephine Porter Institute: https://jpibiodynamics.org/collections/biodynamic-preparations/products/bc-biodynamic-compound-spray-preparation-adapted-from-maria-thuns-barrel-compost-recipe.

8   "About Barrel Compost," Josephine Porter Institute, Floyd, Virginia.

500 and 501. Others, however, reasoned that an additional clay preparation(s) was appropriate. At this time, I'm familiar with three clay preparations—one for summer, one for winter, and one for year-round use.[9]

The winter (fall→spring) horn-clay preparation helps to resist soil nutrient leaching that occurs when plant roots are not actively absorbing them. It strengthens the *ebb* of sap in plants. The summer (spring→fall) variant facilitates the release of nutrients to growing vegetation. It strengthens the *flow* of sap in plants.[10] The year-round (fall→fall) preparation seems to be an early iteration. I have yet to confirm it, but the fall→fall variant seems of particular value to the perennial onions I grow.

Hugo Erbe's #5 Cow Stomach Preparation (HE #5). The essential aim of this preparation is to request assistance from elementals for soil digestion processes. It can be used in making compost or as a soil spray.[11]

Hugo Erbe's #8 Three Kings Preparation (HE #8). Using the gifts of the Magi, Erbe developed this preparation to provide a defined safe zone for elementals to work; a protective boundary that protects against "opposing forces." Three Kings is applied to the periphery of the farm or homestead on Epiphany, preferably at or around 4:00 p.m. EST. It should be used in conjunction with earlier applications of Erbe's #9 Harmonizing Preparation, and *not* used unless the traditional BD preparations 500 to 508 have already been applied. Without prior applications of 500 to 508,

---

9   Lovel, *Quantum Agriculture: Biodynamics and Beyond,* p. 42; Moore, *Stone Age Farming,* pp. 93–94.

10 Lovel, "Advanced Biodynamic Agriculture: How to Make Biodynamics Work Better: A short course prepared for workshops held in the winter of 2003 in Australia and New Zealand," pp. 26–27; Lovel, *Quantum Agriculture: Biodynamics and Beyond,* p. 157.

11 "Hugo Erbe #5-Bovine Rumen Preparation," Josephine Porter Institute (https://jpibiodynamics.org/collections/biodynamic-preparations/products/hugo-erbe-5-bovine-rumen-preparation.

using this is comparable to confining livestock in a fenced pasture that is inferior, depleted, and lacking in life force.[12]

Hugo Erbe's #9 Harmonizing Preparation (HE #9). A companion preparation to HE #8, the Harmonizing Preparation serves as *an expression of gratitude to the elemental beings*. It aids the higher beings (warmth, light, and air) and the lower elemental beings (water and earth) in working together harmoniously for the restoration of fertility to the earth. It should be used in conjunction with HE #8 Three Kings Preparation, and *not* used unless the traditional BD preparations 500 to 508 have already been applied.[13]

> ### About Hugo Erbe
>
> German-born Hugo Erbe (1895–1965) was one of the most insightful and creative pioneers of biodynamics. While some practitioners adhere only to indications and preparations attributed directly to Rudolf Steiner, Erbe adopted a more expansive philosophy. In a document he finished shortly before his death, he wrote, "Steiner never left any doubt that he did not wish his indications to become hardened dogmas, but that they should rather be seen as laying the foundations for a new type of agriculture and as being open to further development and to modification according to circumstances."[14]
>
> Like Steiner, Erbe was clairvoyant and conscious of the role that elementals played in crafting the physical world. He understood that Steiner equated elementals to etheric formative forces, and that the traditional preparations Steiner prescribed worked as nourishment and incentives for the elementals.[15] He carried this deep understanding into his own work, which resulted in twenty-one new preparations.

---

12 "The Three Kings Preparation (HE #8): How It is Made and Applied." Josephine Porter Institute.

13 "Hugo Erbe #9-Harmonizing Preparation," Josephine Porter Institute: https://jpibiodynamics.org/collections/biodynamic-preparations/products/hugo-erbes-9-harmonizing-preparation.

14 Erbe, *Erbe's New Preparations?* p. 22.

15 Ernst Hagemann, "The Elementals as Helpers in Farm and Garden," pp. 31–32, in *Erbe's New Bio-dynamic Preparations*.

Psychotronic Broadcasting

The BD Preparations are physical in that they are composed of tangible materials like quartz, clay, fermented manure, egg shells, or herbs. While these substances are horticulturally beneficial in their own right, we must remember that it is the energy information they carry that makes them powerful and special. When we dynamize Compound Spray Preparation, BD 500, or another such preparation, we transfer that energy information into water and, since many of us use sprayers, we strain out the remains of the physical material so it won't clog the nozzles. The all-important energy information is brought to the soil and plants in water and only the tiniest amount of the physical preparation actually reaches them. When remotely broadcasting the preparations using psychotronics, we accomplish the same thing. We lift the energy information from the physical substance of the preparation and send it to the soils and plants as specified by the witness(es).

I keep a small radionics device operating around the clock, broadcasting select BD Preparations to my homestead. It is an old Rogers instrument with a booster attachment. The Rogers instrument was named for Rogers, Arkansas. As I understand, it was there that a regional radionics group used to meet, probably around the late 1970s and early 1980s. One or more of the members designed, produced, and marketed this instrument. Later, perhaps in the 1990s, the business was taken on by a gentleman in Missouri who made and sold them. Unfortunately, this instrument is no longer being manufactured.

Rogers Radionic Instrument Used to Broadcast to the Whole Homestead

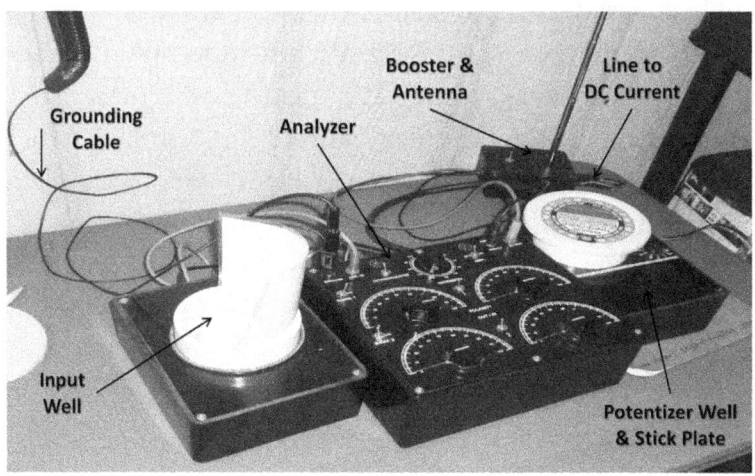

To use a radionics instrument for broadcasting BD Preparations to the whole farm or homestead, I do the following:

- Dowse to determine which preparations ought to be broadcast. If more than one, also dowse to determine whether they can be combined in a single broadcast.
- Insert into the input well:
    - witness(es) of the property, usually photographic or surrogate (see following graphic);
    - the *Engage Card;*
    - either a specimen of the actual preparation(s) or potentized copies.
- Set the rate dials to 00–100.
- Dowse to confirm that it is *safe and advisable* to broadcast.
- Toggle the amplifier switch "on."
- Dowse to determine time required.

When I dowse for time required, I ask, first, whether it is best measured in minutes, hours, or days. Generally, my homestead

Balancing the Whole Farm and Homestead

Surrogate Witness for Property

```
            ← 595 ft →
    N
    ↑        ┌─────────────┐    Property located at
             │             │    210 Frisco Road
             │   Garden    │    Squalor, Missouri
             │             │
             │             │         ⎛ Pond ⎞
             └─────────────┘
  ← 332 ft →                                      ← 332 ft →
    ─ ─ ─ ─ ─ ─ ─ ─ ─ ─ ─ ─ ─ ─ ─ ─ ─ ─
    ┌────┐      ┌─┐ Shop
    │    │      └─┘
    │    │                            ┌────┐
    │    │                            │Pole│
    │House                            │Barn│
    └────┘                            └────┘
              ┌─┐ Storage
              └─┘ Shed   ← 595 ft →
```

preparation broadcasts run one to four days. When finished, I toggle the broadcast switch "off," remove the preparation reagent(s), and dowse my next steps. First, I determine whether another broadcast should be made immediately, and if not, when? If another broadcast is advisable, and it usually is, I dowse for the preparation(s) to use. I also dowse to confirm that witnesses and the preparation reagents I'm using remain suitable.

If you do not have a radionics instrument, consider one of the other psychotronic options described in the chapter on broadcasting. Use pendulum or stick dowsing and try following the same basic steps just presented; make adjustments when appropriate.

The broadcast time required when using alternative instruments might be longer or shorter than that required for a radionics instrument but, in most instances, I suspect it will take longer. Trust your dowsing to make these determinations.

If you have been formally trained in radionics, the aforementioned procedure will be familiar, with one exception—the inclusion of the *Engage Card*. When I leave the *Engage Card* in the instrument well, it functions like a reagent that is added to support the broadcast. If you are uncertain about using the *Card* in this way, dowse first.

The Preparations: Traditional Applications or Psychotronic Broadcasting?

Psychotronic broadcasting adds new dimensions and flexibility to the use of the biodynamic preparations. However, biodynamic practitioners and advocates are right to ask whether it can take the place of traditional applications. In the first paragraph of the previous subsection, I laid out my rationale for using psychotronic broadcasting but, like so many of my explanations, this is part and parcel of my working paradigm. I won't argue that others must agree with me.

According to my still-evolving understanding, the BD Preparations are nourishment for the elementals—or *the etheric formative forces*, if you prefer. Therefore, I trust to my dowsing-assisted communication with co-creative partners; it is these beings that the preparations are intended to support in the first place. I customarily dowse to determine whether an application of the preparations needs to be made in the traditional manner. When indicated, I do so, although I might simultaneously do a psychotronic broadcast to reinforce it.

Radionic Rates for Biodynamic Preparations

Instead of reagent specimens or potencies of the preparations, I prefer to use radionics rates when broadcasting. I've provided these in the following table.

Radionic Rates for Biodynamic Preparations			
Item	*Source**	*Source***	*Source****
500 Horn Manure	52.00–72.00	16.75–05.25	
Pfeiffer BD field spray			61.00–42.35
501 Horn Silica	58.50–89.50	14.00–30.00	
501(F) Horn Silica			80.25–71.25
502 Yarrow Blossoms	25.00–34.00	41.75–92.50	
503 Chamomile	37.00–33.50	15.50–28.75	
504 Stinging Nettle	47.00–34.50	21.25–22.25	
505 Oak Bark	43.00–38.00	23.00–35.50	
506 Dandelion Flowers	52.50–41.50	56.25–78.50	
507 Valerian Flowers	48.50–38.00	10.50–19.50	
508 Horsetail	32.00–86.50	38.00–24.50 23.50–04.50	
Barrel Compost			78.30–21.50
Horn Clay, fall→spring	76.25–47.00		
Horn Clay, spring→fall	50.50–91.00		
Horn Clay, fall→fall			50.25–08.50
HE#5 Cow Stomach			88.00–11.25
HE#8 Three Kings			53.10–56.30
HE#9 Harmonizing			50.25–23.00

Sources: *Hugh Lovel, Quantum Agriculture: Biodynamics and Beyond, Blairsville, GA: Quantum Agriculture, 2014; **Lutie Larsen, Gardening Ratebook, Pleasant Grove, UT: Little Farm Research, 1988; ***developed by the author.

Broadcasting using rates is straightforward, especially if you use a radionics instrument. The following graphic illustrates this. If I were using a radionics template or another of the alternative broadcasters, the arrangement would be similar, except that the rate would be handwritten on a piece of paper or a sticky note, using the protocols advised for intent cards.

Psychotronics and a Biodynamic Garden

Broadcasting BD 500 using a Rate:
- witness card of farm/homestead in the input well
- intent card for working with nature intelligences in input well
- rate for BD 500 set on dial bank
- toggle broadcast switch "ON"
- dowse for time using amplitude dial

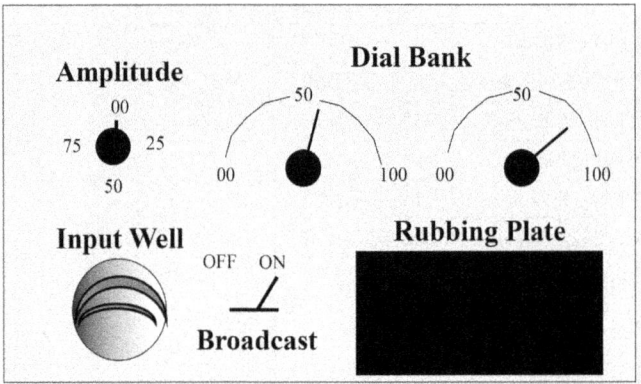

If you have a single dial-bank instrument like the one shown in the graphic, using rates for broadcasting has an obvious limitation: you can broadcast only one preparation at a time. By contrast, the Rogers instrument features two dial-banks allowing me to broadcast two rates simultaneously. I still find that limiting. Clearly, I *could* use reagent specimens of the preparations, or their potencies, and not be limited at all.

I've chosen a third option, however, by creating a full set of preparation intent cards, with each bearing the names and radionics rates. Some examples appear in the following graphic.

I wrote the names and rates near the bottom of each card so that the portion you might contaminate with fingerprints remains above the rim of the well. As with all cards I intend to reuse, I laminate them or store them in glassine envelopes.

Three Examples of BD Prep Intent Cards

BD 500
Horn Manure
52.00-72.00
16.75-05.25

Horn Clay
Fall→Spring
76.25-47.00

Hugo Erbe #8
Three Kings
53.10-23.00

About Steiner, Electricity, and Radionics

Psychotronics is *not* widely accepted in the biodynamics community. There are bound to be many reasons for this, and I'm in no position to judge validity or worth. However, because it is mentioned so frequently, there is one reason I will discuss—the matter of electricity.

Rudolf Steiner was critical of electricity. On January 28, 1923 (about a year and a half prior to delivering the Agriculture Course), Steiner gave a lecture titled "Moral Influences of Light and Unmoral Influences of Electricity" (Dornach, Switzerland).[16] Like many of Steiner's lectures and writings, it requires a depth of understanding I still lack, but here are some salient points:

- Steiner described electricity as *ahrimanic*. Ahriman, in anthroposophic writings, is an adversarial spiritual being, like *Lucifer*, but also his polar opposite. While Lucifer might lead humankind into false spirituality, Ahriman drags it toward a cold materialism.

---

16 Steiner, "Concerning Electricity," lecture in Dornach, Jan. 28, 1923 (*Anthroposophic News Sheet,* no. 23/24 Dornach, Switzerland: The General Anthroposophic Society, June 9, 1940).

- While Lucifer and Ahriman are usually judged as inherently bad, Steiner treats them as polarities that we need to balance to evolve spiritually. This balancing is about becoming mindful and spiritual, while still navigating and stewarding our physical world. Steiner actually railed against hypocrites who spoke forcefully against ahrimanic electricity at his lectures, but did not hesitate to ride home on electric trolley cars.

Until now, I've not discussed the role of electricity in psychotronics. Though it is common for many radionics instruments to be powered by electricity, it is generally only computerized units that "require" it for all functions. The Rogers instrument does not require electricity for dowsing, for analysis, or for most balancing. Customarily, I *never* use electricity when broadcasting to myself, other people, pets, or individual plants. However, when broadcasting to fields, garden, or homestead, I appreciate the added boost electricity provides, and the shortened time required.

In the past year or so, I've included my co-creative partners in decisions regarding electricity. I'm discovering that electricity is recommended for some broadcasts to my homestead, but not for others. I've found no clear pattern as yet that explains this.

A Few Words about Sanctuaries

Biodynamics supports the creation of natural biodiversity areas within farms and homesteads. Where possible, growers are encouraged to set about ten percent of their land aside and allow it to evolve naturally.[17] Being part of nature ourselves, we are encouraged to enjoy these landscapes and resist the human urge to mess with them.

In writing about *Elemental Annexes*, Machaelle Small Wright describes something more specific. Like the biodynamic concept, she describes landscape areas that are left to nature and not managed. However, she suggests even stiffer constraints on human activity:

---

17 Jim Fullmer and Janet Gamble, "Organic Agriculture and Biodynamic Methods: A Workshop Program within the Organic University Portion of the 2009 MOSES Conference in La Crosse, WI," Feb. 26, 2009.

> Designating an area for a nature sanctuary means that we are committed to not entering, altering, adjusting, "cleaning" or "beautifying" the spot in any way. It is not to double as a sitting or meditation area or a play area for children. However, pets and wild animals fall under the category of nature and, when they enter or pass through a nature sanctuary, it is not considered an intrusion.[18]

Small Wright goes on to say that it is fine to observe what happens within sanctuaries, as long as we don't enter them. I dowsed the location for a sanctuary shortly after we purchased our homestead and it has remained untouched for about twenty-five years. Sometimes I purposely avoid even looking into my sanctuary because my German heritage favors ordered and well-kept spaces. I have to fight the urge to clean up dead branches and do other obsessive things.

---

18 Small Wright, *The Perelandra Garden Workbook*, p. 31.

13

Balancing the Garden and Field

In this chapter, I narrow my focus to the garden level. The garden is nested within my homestead, so it receives the consideration and benefits already provided to the wider property. However, the land I cultivate for food and other purposes has additional needs. For someone working on the larger-scale farm, many of these techniques and ideas apply to individual fields.

The psychotronic protocols I use on the garden level do not change much from those I used in the previous chapter. The *Engage Card* is employed the same as before. I do, however, replace witnesses for my whole homestead with ones for the garden or a specific field. As before, these witnesses can be either a photo or a surrogate, like the one shown in the next graphic.

I should explain that it is not necessary to remove the whole farm or whole homestead witnesses. By simply adding the garden or field witnesses, your focus is properly narrowed; I only suggest replacing them because input wells on many radionics instruments are small and limiting witnesses frees up some workspace.

Gardening Methods

As with most things, gardening can be done in many different ways. Therefore, growers need to make lots of decisions about lots of issues:

Balancing the Garden and Field

Surrogate Witness for Garden or Field

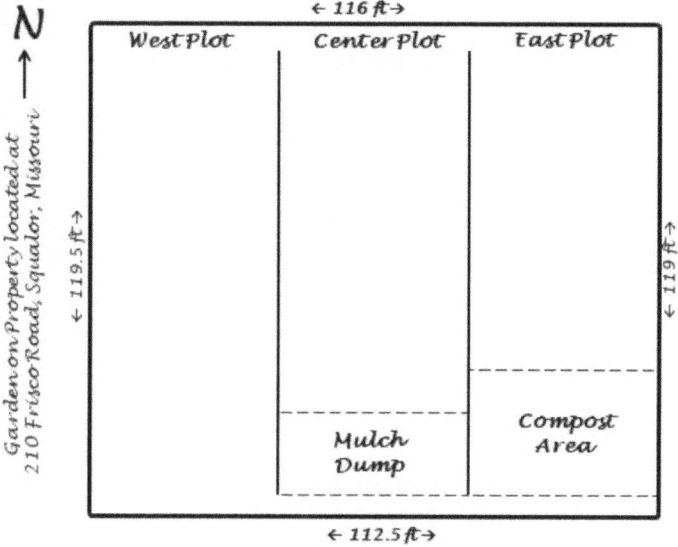

- Should I use clean culture? This is what most gardeners do. They plow, disk, rototill, or dig the soil before planting. They keep the soil between plants bare and clear weeds by cultivating, hoeing, and pulling them.
- How about mulching? As a full or partial alternative to clean culture, some growers use straw, rotted hay, shredded leaves, grass clippings, or wood chips to cover the soil and suppress weeds.
- What about plastic mulches? Plastic mulches also suppress weeds and combine well with clean culture.
- Composting? Composting is common to many gardening systems, but is most associated with organic management. Compost is excellent for soil-building, but you can also apply manure, vegetation, and food wastes in raw, non-composted forms.
- Raised beds or planting ridges? Many growers prefer these on wetter, heavier soils.

Psychotronics and a Biodynamic Garden

- Insecticides and other pesticides? The choices are usually between conventional synthetic pesticides and organically approved materials. Whether you use them or not, pesticides should remain your last option.

The choices we make on these and other matters affect our gardening success, the quality of produce we harvest, and the impact we have on the wider environment. Perhaps we are not the most eco-friendly gardeners in our neighborhood, but dowsing with the aid of co-creative partners will, with time, nudge us toward means and practices that are in better harmony with nature.

Machaelle Small Wright recounts that she began gardening in a very conventional manner—using cultivation, straight rows, even Sevin®, a synthetic insecticide.[1] As she partnered with nature intelligences, she was guided to lay out the garden in a circle. She eliminated much of her digging and hoeing and adopted mulching—using the methods taught by Ruth Stout.[2]

A Word on Ruth Stout

Ruth Stout (1884–1980) was a pioneering gardener and researcher who promoted year-round, deep-mulch gardening. She authored many articles that were published in *Rodale's Organic Gardening and Farming* magazine. She also wrote several how-to books that detail her mulching methodology, including the 1955 classic *How to have a Green Thumb without an Aching Back*.

---

1 Small Wright, *The Perelandra Garden Workbook*.
2 Ibid., pp. 4, 83–84.

A Bio-extensive System

When I designed the Cannon Horticulture Project for the Kerr Center, I had the opportunity to try a management system I'd thought about for several years—one I believed would suppress weeds, control many soil diseases and insect pests, reduce labor, build soil, and be self-sufficient in nitrogen. It is called a bio-*extensive* system.[3] Bio-extensive systems use more land than most gardening and farming approaches, but the extra field space used is a trade-off for those many benefits.

The dominant features of Kerr Center's bio-extensive system were planned crop rotations and the creative use of cover crops. To this we added organic no-till practices, beneficial insect habitats, and living mulches.[4]

I brought many elements of bio-extensive management to my biodynamic garden. They have fit very well so far, but more time is needed for evaluation.

What Crops to Plant

It is spring as I complete these final chapters. They reflect the state of my planning and earliest 2020 gardening activities. In many cases, I'm describing procedures without knowing the end results. Still, most everything I detail emerges from seasons and years of trial and error.

When deciding what to plant in 2020, I made more lists. (Are you surprised?) I have one for vegetables, one for herbs, and another for companion plants. For my purposes, companion plants are a mishmash of flowers, beneficial habitat plants, pest-repelling species, and anything else that isn't obviously a vegetable or herb. I

---

3   Not to be confused with bio-*in*tensive systems, which are designed to produce as much as possible in small areas.

4   For detailed information on Kerr Centers Bio-extensive system, see: Kuepper, *Market Farming with Rotations and Cover Crops: An Organic Bio-extensive System* (https://kerrcenter.com/publication/market-farming-with-rotations-and-cover-crops-an-organic-bio-extensive-system); also, Kuepper, *Organic Bio-extensive Management Revisited*.

started each list with crops I'd be interested in growing. (They didn't need to be favorites of mine.) The lists appear below.

Lists of Annual Plants for the 2020 Garden

Vegetables		Herbs		Plants	
Radish		Basil		Sunflower	√
Lettuce	√	Parsley		Castor bean	
Collards		Mint		Marigolds	√
Irish potatoes	√	Coriander		Hollyhocks	
Beets		Dill	√	Milkweed	
Chard	√	Garlic		Nasturtiums	
Carrots		Peppermint		Yarrow	
Cabbage	√			Sesame	
Onions				Zinnias	√
Sweet corn					
Green beans					
Dry beans					
Okra	√				
Tomatoes	√				
Cucumbers					
Summer squash	√				
Winter squash					
Sweet potatoes	√				
Watermelons					
Cantaloupe					
Southern peas	√				
Peppers	√				
Eggplant					
Pumpkins	√				
Broccoli	√				

I have a bad tendency over-do; I set out to accomplish more than I should and get overwhelmed. To curb this, I dowsed to find those crops that my co-creative partners felt were most important to grow in 2020. I placed a check (√) next to those recommended.

Next, I asked, "Are there additional crops or plants I should plant in the garden this season?" Receiving a "yes" answer, I Scan-Dowsed available lists of plants to discover that my co-creative partners also wanted me to grow pumpkins, broccoli, and zinnias.

What Varieties to Plant

Now that I know which crops to grow, I need to know the most suitable varieties. Since there are huge numbers of cultivars available for most crops, you might do some advance research to narrow your choices. Earlier, I suggested speaking with local growers and consulting the Cooperative Extension Service. Each state extension publishes the names of recommended varieties. Arkansas, for example, recommends cultivars in each of its individual crop fact sheets.[5] However, while I find this information helpful, I'm not constrained by it.

I start my search by creating another list of favorite in-print and online seed catalogs. I'm particularly interested in seed companies that provide heirloom varieties. Because ordering commercial seeds can be expensive, don't hesitate to include bulk seeds from the local farmers' co-op or packets from seed racks in local stores. It will certainly complicate research and dowsing procedures if you try to follow my method, but the extra effort might save you some money. I also make another list of any surplus seed left over from previous seasons, as well as any I saved from crops I'd grown.

---

5   Craig R. Andersen, "Guide to Vegetable Culture," Home Gardening Series Fact Sheet FSA 6001, Fayetteville, AR: Division of Agriculture, University of Arkansas, 2009, p. 1.

About Heirloom Varieties

I prefer *heirloom* (also known as *heritage*) varieties. They are closer to their wild ancestors and retain the genetic diversity and expression that contemporary plant breeding often discards. These include unique flavors, colors, higher nutrient content, and better adaptability to organically managed agroecosystems. Growing and perpetuating heritage varieties counters the ongoing loss of food crop biodiversity—another impending ecological crisis.

The decline in crop nutrition is a further concern. I wrote the following a few years ago[6]:

> Studies done in Canada,[7] Great Britain,[8] and the United States found historical reductions in nutrient density of modern fruits and vegetables.
>
> Two studies were conducted in the U.S., both using data collected and published by the USDA's Nutrient Data Lab. The first study, conducted by a nutritionist at the Kushi Institute in Massachusetts, studied nutrient changes from 1975 through 1997. It found vitamin and mineral content declined as much as 25–50% in both fruits and vegetables.[9] The second study, done by a team at the University of Texas at Austin, studied 43 garden crops over a period from 1950 to 1999. Looking at average changes, they found "reliable declines" in six nutrients—protein, calcium, phosphorus, iron, riboflavin, and ascorbic acid. The differences ranged from 6% for protein to 38% for riboflavin.[10]

It is true that the declines in nutritive value can be due to other factors including soil depletion and the higher levels of atmospheric carbon

---

6 Kuepper, *Heirloom Vegetables, Genetic Diversity, and the Pursuit of Food Security*, pp. 1–2 (https://kerrcenter.com/publication/heirloom-vegetables-genetic-diversity-and-the-pursuit-of-food-security).

7 André Picard, "Today's foods lack yesterday's nutrition," Bell Globemedia Interactive (http://www.theglobeandmail.com/special/food/wxfood.html).

8 "Mineral Depletion in Food," *Organic Voice*, 9(2): 21, 2005.

9 Alex Jack, "Nutrition under Siege," *One Peaceful World Journal* 34(1), 1998, pp. 7–9.

10 Donald R. Davis, Melvin D. Epp, and Hugh D. Riordan, "Changes in USDA Dood Composition Data for 43 Garden Crops, 1950 to 1999," *Journal of the American College of Nutrition*, 23(6), 2004, pp. 669–682.

Balancing the Garden and Field

dioxide due to climate change.[11] However, there is consensus that much—perhaps most—of the decline is an unintended consequence of modern plant-breeding strategies meant to increase yield, ease of shipping, and other factors more suited to industrial agriculture than to food quality and consumer health.

It is true that heritage cultivars generally yield less than contemporary ones. They also tend to do poorly on soils that have been abused under chemical management. Therefore, commercial growers are reluctant to raise heirlooms unless they see a strong market demand for them.

Lastly, heirloom varieties are non-hybrids; you can save and replant their seed. Seed saving is encouraged in biodynamic agriculture; it encourages self-sufficiency. And, according to BD proponents, seed raised and replanted on the same farm or garden will, as time goes by, form a positive relationship between the land and the grower.

I created the blank form that follows as an aid for dowsing crop varieties. In the first column, I enter the name of one of the crops, in no particular order. I dowse to see if more than one variety should be grown and if so, how many.

To find the preferred cultivars, I then dowse my seed source lists to find "the first source for one of the optimal varieties I should grow." You might want to start with the list featuring seeds you already have in hand. If the optimal cultivar(s) are not among your carry-over or saved seed, move to the list of catalogs and online sources. Proceed to the pages where the crops' varieties are listed and either Scan-Dowse or YES/NO dowse to find the ones recommended. Enter the name in the second column. Repeat the process until you've found all of the recommended cultivars for the crop.

---

11   Lydia Noyes, "Climate Change and Plant Nutrient Levels," *Mother Earth News*, April/May, 2018 (https://www.motherearthnews.com/natural-health/nutrition/climate-change-plant-nutrient-levelszmoz18amzphe).

Varietal Selection for Annual Plants in the 2020 Garden		
Plant/Crop	Selected Varieties	Preferred Source(s)

To better illustrate, I show the completed example from my 2020 plans. The first crop I chose to dowse was lettuce. I discovered that the optimal source was Pinetree Seeds, a company that sends their annual catalogs to me. I turned to the pages describing lettuce varieties and quickly dowsed one named "Summertime."

Balancing the Garden and Field

Varietal Selection for Annual Plants in the 2020 Garden		
Plant/Crop	Selected Varieties	Preferred Source(s)
Lettuce	Summertime	Pinetree Seeds
Irish potatoes	King Harry	Saved seed stock
Chard (1)	Lucullus	Sow True Seed
Chard (2)	Rainbow Blend	Sow True Seed
Cabbage (1)	Express Red	Peaceful Valley Farm
Cabbage (2)	Golden Acre	Peaceful Valley Farm
Cabbage (3)	Glory of Enkhuizen	Baker Creek Heirloom
Okra (1)	Star of David	Baker Creek Heirloom
Okra (2)	Jing Orange	Baker Creek Heirloom
Tomatoes (1)	Granny Cantrell's German	Saved seed stock
Tomatoes (2)	Paul Robeson	Baker Creek Heirloom
Tomatoes (3)	Martian Giant Slicer	Southern Exposure
Summer squash	Yellow Crookneck	Saved seed stock
Sweet potatoes	Georgia Jet	Saved seed stock
Southern peas (1)	Big Red Ripper	Southern Exposure
Southern peas (2)	Piggott Pea	Southern Exposure
Peppers (1)	Sweet Purple Beauty	Saved seed stock
Peppers (2)	California Wonder	Baker Creek Heirloom
Pumpkins	Creek Indian Pumpkin	Saved seed stock
Broccoli	Calabrese	Southern Exposure
Dill (1)	Mammoth	Saved seed stock
Dill (2)	Bouquet	Saved seed stock
Sunflower (1)	Mammoth Gray Stipe	Saved seed stock
Sunflower (2)	Snacker	Botanical Interests
Marigolds	Colossus Red Gold Bicolor	Baker Creek Heirloom
Zinnias	Queen Lime Red	Baker Creek Heirloom

Dowsing Row Length and Plant Spacing

Before purchasing seed, you should know how much you'll need. This leads to the next step in my process, which begins by dowsing row lengths and plant spacing. To aid in this, I created yet another blank form.

Psychotronics and a Biodynamic Garden

Row Length – Number of Plants – Plant Spacing in the 2020 Garden			
Crop & Varieties	Feet of Row	Number of Plants	Final Spacing of Plants

The first column is self-explanatory. It lists the varietal names dowsed and shown in column two of the previous table. For column two of *this* table, we dowse the number of row-feet of each cultivar that should be planted. Observe that this is done primarily for crops that are direct-seeded into the garden, such as radishes, carrots, beets, chard, okra, beans, and peas. Once I know this information,

Balancing the Garden and Field

I can return to my seed sources and/or Extension literature to find the amount of seed required.

Column three is similar to column two, but mainly addresses crops to be transplanted, such as tomatoes, peppers, eggplants, cabbages, and broccoli. By dowsing how many transplants you'll need, you know how many to purchase, or you can estimate the seed needed to grow your own. I also prefer to dowse for the number of plants needed when I grow widely spaced crops like squash and pumpkins, even if I seed them directly.

I save the last column for final plant spacing. I dowse this information to tell me how far apart to set my transplants. For direct-seeded crops, I'm dowsing to find how thick my final stand should be. (It is customary to seed heavier than required and *thin* the crop to the desired stand once plants are established.) As I do in many such circumstances, I first learn the range of recommended plant spacings from seed sources and/or Extension literature before dowsing. A table is provided to illustrate this.

When to Plant

Cooperative Extension provides good guidance for safe planting and transplanting dates. It is typically a bit conservative, which is good. You are less likely to be surprised by early and late freezes that kill or damage seedlings, flowers, and other tender vegetation.

With a range of recommended planting dates in hand, I turn next to gardening almanacs. These resources provide the optimal dates for planting and tending crops based on their *elemental affinity*. The table on page 165 helps to illustrate this.

For a practical example, I will plant Creek Indian pumpkin in 2020, as I have for several years. I first acquired seed from Oklahoma members of the Creek tribe around 2009. It is a hardy and productive variety that naturally resists squash vine borer—a common and destructive pest in Arkansas and Oklahoma. Resembling "Dickinson," it also belongs to the species *Moschata*.

Psychotronics and a Biodynamic Garden

Row Length / Number of Plants / Plant Spacing in the 2020 Garden			
Crop and Varieties	Feet of Row	Number of Plants	Final Spacing of Plants
Summertime lettuce	2	*	9 in.
King Harry potatoes	30	*	14 in.
Lucullus chard	4	*	5 in.
Rainbow Blend chard	2	*	3 in.
Express Red cabbage	*	4	18 in.
Golden Acre cabbage	*	2	18 in.
Glory of Enkhuizen cabbage	*	2	18 in.
Star of David okra	20	*	26 in.
Jing Orange okra	20	*	24 in.
Granny Cantrell's tomato	*	4	20 in.
Paul Robeson tomato	*	4	20 in.
Martian Giant tomato	*	4	20 in.
Yellow Crookneck squash	*	4	20 in.
Georgia Jet sweet potato	36	*	12 in.
Big Red Ripper peas	20	*	6 in.
Piggott peas	20	*	6 in.
Sweet Purple peppers	*	4	18 in.
California Wonder peppers	*	6	18 in.
Creek Indian pumpkin	*	10	48 in.
Calabrese broccoli	*	4	18 in.
Mammoth dill	2	*	7 in.
Bouquet dill	2	*	8 in.
Mammoth sunflower	8	*	16 in.
Snacker sunflower	8	*	16 in.
Colossus marigolds	2	*	9 in.
Queen Lime Red zinnias	4	*	10 in.

In northwest Arkansas, one can plant pumpkins beginning in April. However, by waiting until July, the population of squash bugs will have declined naturally. Furthermore, the pumpkin fruits will mature just before traditional harvest. I will have these parameters in mind when I consult my preferred planting guide, *The Maria*

Thun Biodynamic Almanac: North American Edition, by Matthias Thun. I look to find the optimal planting dates for fruiting crops in July 2020 and find they occur July 4–5, 13–15, and 22–24.

Crops by Primary Elemental Affinity and Almanac Grouping			
Root Crops	Leaf Crops	Flower Crops	Fruit Crops
Affinity to Earth	*Affinity to Water*	*Affinity to Air*	*Affinity to Fire*
Artichokes, Jerusalem	Asparagus	Artichokes	Beans
Beets	Brussels sprouts	Broccoli	Corn
Carrots	Cabbage	Cauliflower	Cucumber
Celeriac	Celery		Eggplant
Onions, bulbs	Chard	Decorative flowers	Melons
Potatoes, Irish	Kale	Beneficial habitats	Peas
Radish	Kohlrabi		Peppers
Turnips for roots	Leek		Pumpkins
Sweet potatoes	Lettuce		Squash
	Onions, green		Tomato
	Rhubarb		Tomatillo
	Spinach		Watermelons
	Turnips for greens		
			Grain crops
	Lawn grasses		Berries
	Forage crops		Grapes
	Sorghum for syrup		Fruit trees
	Most cover crops		Nut trees

Next I dowse to find which range of dates is optimal, and do my best to schedule the planting accordingly. I've discovered that there is benefit from even more precise timing—trying to nail down the best day and time of day—but few growers have the time and motivation to do that.

Where to Plant

If you grow your crops in a planned rotation, the following discussion might be of limited use. If you have no such strategy, however, it might be worth the effort. By working with co-creative partners, I feel I'm putting my crops in the best places and getting benefits I'd expect from both planned crop rotation and companion planting.

My current procedure requires an additional hand-drawn map of the garden. This map should not be confused with the one you may already be using as a surrogate witness, though you will use both for this procedure.

You won't be sizing it to fit into an instrument, so make this map large enough so you can mark the locations of crops as you dowse them. The map must include all features that you and your co-creative partners need to plan around, such as established perennials. I've included X and Y axes, marked off in feet. Such detail is optional unless you require that level of precision or share my obsessive German heritage. Likewise, I plant most everything in straight, single rows and mulch the inter-rows. This is better for making observations and the additional experimentation I do.

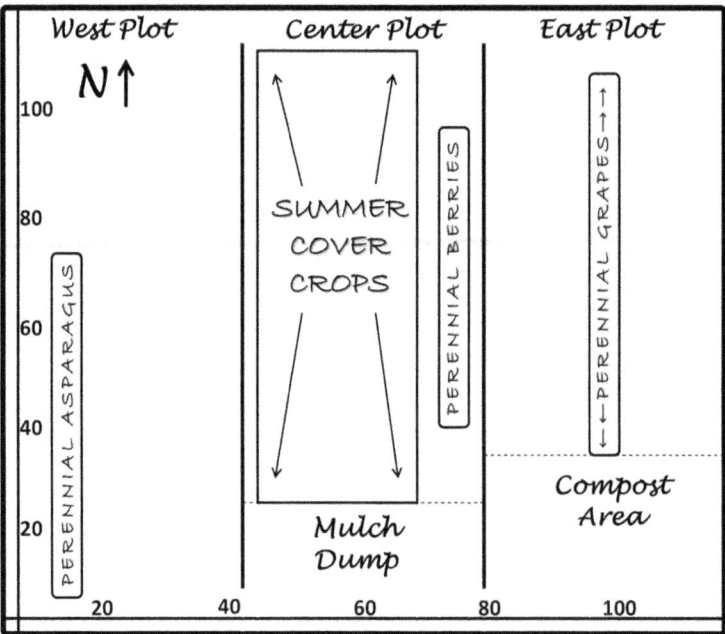

Balancing the Garden and Field

To find the optimal location for a specific crop, I Scan-Dowse the horizontal axis from left to right, seeking the optimal x-coordinate. Next, I Scan-Dowse the vertical axis from bottom to top, seeking the optimum y-coordinate. I interpret the x–y intersection as the beginning of each row going north. Since I already know the number of plants and/or the row-feet for crop, I can sketch a fairly accurate entry onto the map. I show this on the next map, where I've entered the locations for a few of the crops.

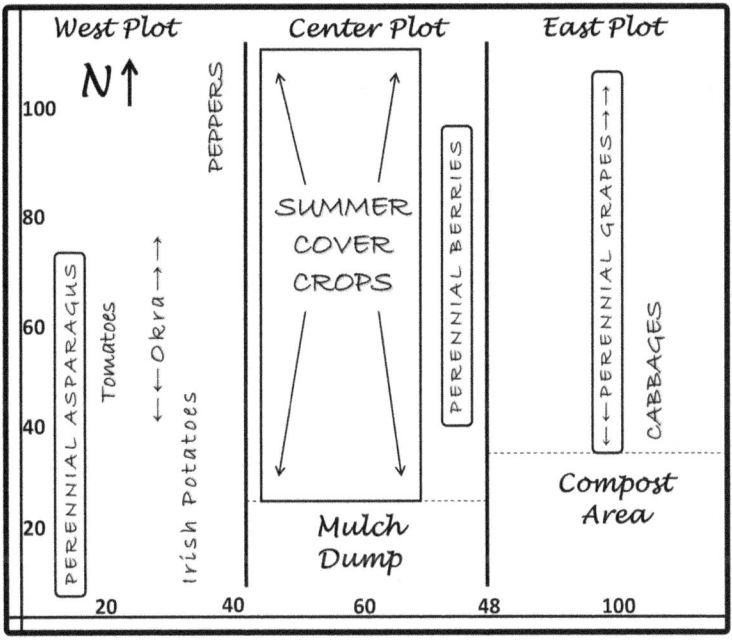

Remember to Retain Seed Witnesses

Now that you know what, when, and where to plant, it remains for you to get out there and actually do the gardening. However, if you have not already retained witnesses, you will want to do so before you find you've planted every last seed. It takes time to get in the habit, but you will find it's easy to draw a few seeds from each packet or bag, place them in a glassine envelope or lead-free glass tube, label them, and store them as witnesses.

Making Compost: An Experimental Approach

I hate admitting this, but I'm not a particularly talented gardener. One of my weakest skills is composting. I have, on occasion, made a few decent batches, but it's been mostly hit or miss. So, I'm drawing on the guidance of co-creative beings to improve my performance. I want to develop an approach to composting that uses psychotronics and co-creative wisdom in ways similar to those I've already described in this book.

Since it is already early 2020, and there is much to do this spring, I foresee little time to implement the procedure I've put together. So, unlike the other techniques in this book, all of which arose from one or more seasons of trial and observation, this one is not, as yet, ground-tested. I'm describing a procedure I *will* be trying out.

My strategy begins with several assumptions:

- Major feedstock materials will all be acquired on-site or locally.
- Any commercial feedstocks and amendments will be low-cost and affordable.
- Biodynamic compost preparations will be used in the traditional manner.
- Internal temperature of the pile will be monitored.
- Any turning of the pile and watering will be determined by dowsing with co-creative assistance.

The following graphic lists the feedstocks and amendments that meet my criteria for cost and availability.

Balancing the Garden and Field

Lists of Available Compost

Feedstocks	Amendments
Broken twigs	Aglime
Leaf & pine needle mix (not ground)	Alfalfa meal
Half-year-old lawn clippings	Feather meal
Wood chips	Bone meal
Partially composted vegetation	Soft rock phosphate
Well-rotted pine needles	Gypsum
Commercial bagged soil	Builders' sand
Commercial bagged cow manure	
Commercial bagged composted poultry litter	

There are lots of ways to construct compost piles. I'm designing mine based on sequenced layers that repeat until the pile reaches a satisfactory height and breadth. The main things I dowse are to determine the feedstocks to use, when to layer each as I build the pile, and the desired thickness of each layer. I use the witness(es) for my garden and the Engage Card. I keep the list of compost feedstocks and amendments at hand for dowsing.

Because this procedure is still theoretical, I won't walk through the steps in detail, but my presumed results are shown in the following table, including each feedstock, where it should be layered and the depth of each layer. Following that is a graphic illustrating how things might appear in an actual compost pile. You might be skeptical whether wood chips will make a suitable cover layer. I expect they would work poorly were my pile shaped in the manner depicted. However, I will actually use large cylindrical plastic compost bins and expect the chips to work well in those.

Materials, Layering, and Amounts: First 2020 Compost Pile

Feedstock or amendment	Layer	Amount per Layer
Wood chips	cover	3 inches
Bagged soil	6th layer, repeat	10 pounds
Gypsum	5th layer, repeat	1/8 cup
Rotted pine needles	4th layer, repeat	2 inches
Poultry litter compost	3rd layer, repeat	1 inch
Partially composted vegetation	2nd layer, repeat	1 inch
Leaves and pine needles	1st layer, repeat	3 inches
Broken twigs	Base	1 inch

Experimental Compost Windrow 2020

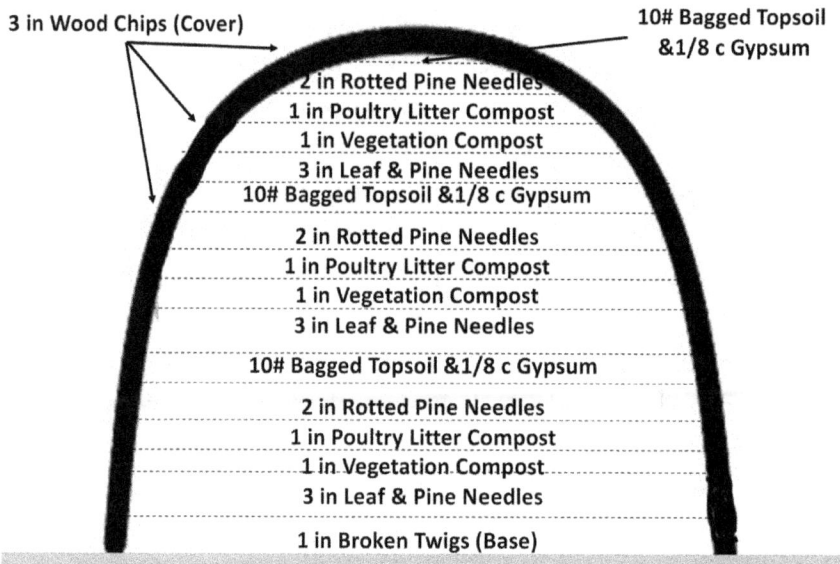

14

Balancing Individual Crops and Plants

Now I am shifting focus to the innermost level, or "nest," in my management scheme. Here I address the procedures that deal with individual crops or plants within my garden. If I were working on a large farm scale, these techniques might be more useful with crop fields than those I described in the preceding chapter.

I continue to use the garden witness(es) and the *Engage Card* as in the previous chapter. Now, however, I add specific crop or plant witnesses. If I've failed to retain seed witnesses, I obtain a photo and/or traditional leaf or stem specimens.

For better organization, I gather these witnesses into root, leaf, flower, and fruit crop categories, as explained in the previous chapter. My almanac is important now, so I keep it close by. As I mentioned earlier, I prefer to use Matthias Thun's annual guide,[1] but you can use whatever is most helpful to you.

Tending the Crops

The almanac is my prime source for discovering which dates are best for tending to root, leaf, flowering, or fruiting plants. I plan well ahead. I keep a daily planner and enter activity dates at least two months in advance so that I'm well-prepared when optimal transplanting, fertilizing, or pruning dates come around.

---

1   *The Maria Thun Biodynamic Almanac: North American Edition.*

Psychotronics and a Biodynamic Garden

For example, let's assume it is mid-June of 2020 and we're in a fruiting period (the full days of June 16–17). This will be a good time to tend to my Granny Cantrell tomatoes. I pull out my Granny Cantrell witness and combine it with the whole garden witness and the *Engage Card*. I then dowse which, if any, of the following tasks should be done at this time:

- Weeding
- Mulching
- Watering
- Training
- Pruning
- Applying BD Preparations
    » Traditional Application
    » Psychotronic Broadcast
- Fertilizing
    » Dry Sidedress
    » Drench
    » Foliar Spray
    » Psychotronic Broadcast
- Controlling Diseases
    » Destroying Plant(s)
    » Pesticide(s)
    » Psychotronic Broadcast
- Controlling Insect Pests
    » Hand-picking
    » Row Covers
    » Beneficial Insect Release
    » Pesticide(s)
    » Psychotronic Broadcast

Balancing Individual Crops and Plants

Weeding, Mulching, and Watering?

These activities are largely self-explanatory. If mulch is indicated, I might want to dowse my options. Should I use pine needles, grass clippings, or wood chips? Plastic is usually not advised once a crop is already planted or transplanted, so I don't consider it an option.

Understand, however, if your crop is stressed from lack of water, serious pest infestation, or disease, don't look to your almanac for when to take action! If plants need immediate attention, you need to do something right away.

Training and Pruning?

Among my annual fruiting crops, tomatoes are the only ones that I train. If you are in a humid state like Arkansas, you will probably want to train your tomatoes, too. Allowing them to run along the ground invites disease, especially if you use clean culture.

Pruning to one or two stems is popular with some gardeners, but it is largely optional. I've never really bothered to do it.[2] Both training and pruning are ongoing and time-sensitive activities. You probably shouldn't wait for optimal fruiting dates to do them.

Biodynamic Preparations?

Individual crops sometimes benefit from targeted applications of a BD Preparation. Hugh Lovel discussed this in a 2018 webinar,[3] when he pointed out the value of using BD 505—the Oak bark preparation—following heavy rains. Heavy rainfall, especially after a dry spell, can increase nitrification in the soil and lush watery plant growth, leading to pest insect infestations and fungal diseases. BD 505 helps to sequester nitrates and reverses nitrification.

---

[2] Kathryn Fontenot, "How to Properly Prune Your Tomatoes," AgCenter, Louisiana State University (https://www.lsuagcenter.com/topics/lawn_garden/home_gardening/vegetables/home_garden_crops/pruningtomatoes).

[3] Hugh Lovel, "How Biodynamic Preparations Work" (webinar), Biodynamic Association of North America, Dec. 14, 2018.

Fertilization?

When I began using radionics in 1986, I focused on compounding fertilizers for blueberries and other horticultural crops. This strategy was enormously effective, as I recounted elsewhere. I used classic radionics methods and detailed them in *Plants, Soils, Earth Energy, and Radionics*.[4] I won't elaborate further in their regard.

My procedures these days are less exacting. If I dowse that fertilizer is needed, I go on to dowse which products or materials are most advisable. I also consider supportive aids like sugars, humates, surfactants, microbial inoculants, and vitamins that might also be beneficial. I intentionally limit myself to items I already have on hand, and to those I can afford. The organic farming and gardening marketplace, like the conventional one, is overrun with high-priced fertilizers and amendments. While many are quite good, they can quickly drain your bank account if you don't use common sense. Besides, good organic and biodynamic practices provide most of the benefits commercial products promise.

The physical–chemical nature of the fertilizer you dowse is likely to limit how you apply it. If it is dry and insoluble (such as greensand, hard rock phosphate, or kelp meal), your sole option is to sidedress it. If the fertilizer materials are liquid or soluble in water, sidedressing is one of three options. You might also drench each plant or spray the fertilizer onto the leaves.

If the material is dry, sidedressing means laying a band of it alongside the crop row or around the base of the plant; if it is liquid, it would be poured or dribbled. Sometimes an effort is made to work it into the soil; sometimes not. To determine how much to apply, first check the label directions, then dowse the precise amount you should use.

Drenching differs from sidedressing in that you use enough additional water to carry the fertilizer down into the root zone. It is

---

4   Kuepper, *Plants, Soils, Earth Energy, and Radionics*.

Balancing Individual Crops and Plants

obviously limited to liquid and soluble fertilizers, and is best suited to large individual crop plants like tomatoes, squash, and pumpkins.

When drenching, I first dowse the desired fertilizers; then I dowse how much of each is required to treat *all* of the plants. My garden is small, so I use a watering can for drenching. I dowse for the number of cans of liquid I need to ensure even distribution of the fertilizer to an optimal root depth. If, for example, I dowsed that a pint of liquid fertilizer was needed for my twelve tomato plants, and four cans of total liquid were required to drench them, I would mix four ounces of fertilizer with water in each can.

I've always had a great interest in foliar fertilization. In the 1980s, it played a huge role in recovering Kerr Center's blueberry planting in Oklahoma. However, my scale of gardening today is much smaller, and foliar feeding often seems less practical. Still, I use it occasionally.

Foliar fertilizing is more exacting than either sidedressing or drenching, but it is not complicated. After dowsing which fertilizer(s) should be applied, I go on to dowse how much should be applied to the entire crop. I simply add that amount to slightly more water than I think I'll need and spray until it is used up. Were I working on a larger scale, I would return to the more precise methodology I described in *Plants, Soils, Earth Energy, and Radionics*. Here are some rules of thumb for using foliar sprays, no matter what the scale of application:

- Use only clean water. If spraying a microbial inoculant and using chlorinated water, hold it for at least 24 hours to allow the chlorine to gas off; aeration speeds the process. Alternatively, putting a few drops of humic acid extract into the water complexes and neutralizes the chlorine. There are commercial filters that can dechlorinate water, but I've little experience with them. No matter the method, always dowse to determine whether your water is suited to its intended use.

- Spray in the cool of early morning or late evening when plant stomates are open and absorption increases. Spraying on a hot sunny afternoon is a waste of time.
- Avoid windy days or much of your spray will land elsewhere.[5]
- Sometimes the benefits of a fertilizer or amendment can be achieved through psychotronic broadcasting. You can make a fertilizer reagent for broadcasting by placing the material(s) in lead-free glass; glassine envelopes will also work if the substance is dry.

When broadcasting fertilizers, I prefer to use radionic rates, simply because they are easier to work with. Often, though, rates for materials you want to use have not been developed. In such cases, you either need to make up a reagent and, if you want, scan your own rate for future use. A list of useful fertilizer and amendment rates follows.

Not included in the fertilizer rate list are homeopathic fertilizers based on *Schuessler's Cell Salts*. The first commercial formulation I knew of was called *Cosmo*.[6] It was formulated and promoted by the English radionics pioneer Bruce Copen (1923–1998) and is briefly discussed in his 1980 book *Electronic Homeopathy for Plants*.[7] To my knowledge, Cosmo is no longer being formulated. Fortunately, T. Galen Hieronymus, one of his coworkers or a cooperator, scanned and recorded a rate for Cosmo: 26.50–45.50.[8] I have used this rate quite often.

---

5 Ibid., pp. 132–134.
6 Peter Lindemann, "Cosmo Fertilizer," *The Journal of Borderland Research*, July/Aug. 1989, p. 27.
7 Copen, *Electronic Homeopathy for Plants*, pp. 28–29.
8 Hieronymus, *Cosmiculture*, p. 64.

Balancing Individual Crops and Plants

Radionic Fertilizer Rates				
Item	Rate	Source 1	Source 2	Source 3
Acetic acid	17.50–28.50	*		
Alfalfa meal	35.00–48.75	*		
Apple pomace	27.00–23.80	*		
Basic H	48.50–71.50		*	*
Basalt	37.50–18.25	*		
Blood meal	42.75–54.25	*	*	*
Bone meal	18.00–33.75	*	*	*
Borax	24.00–52.50	*	*	
Clay, bentonite	30.25–34.75			*
Charcoal	42.50–66.75	*		
Coffee grounds	35.75–39.50	*		
Copper sulfate	54.50–75.75	*	*	*
Cottonseed meal	31.00–44.50		*	*
Epsom salts	60.75–47.25			*
Evergreen needles	34.50–13.25	*		
Feather meal	21.00–37.25	*		
Grape pomace	35.50–52.50	*		
Granite dust	38.00–53.50	*	*	*
Greensand, NJ	46.50–76.75	*	*	*
Gibberellic acid	50.50–62.50		*	*
Guano	44.50–12.25	*		
Gypsum	40.75–41.50	*	*	*
Honey	14.50–25.50	*		
Hoof and horn meal	77.50–81.00	*	*	*
Hop manure	28.50–30.50	*		
Humic acid	20.25–26.75	*	*	*
Hydrogen peroxide	05.75–30.50	*		
Iron sulfate	60.75–47.25			*
Kelp meal	50.00–66.25		*	*
Leaf mold	43.00–94.00	*	*	*
Leather meal	41.25–93.25		*	*
Lignite, soluble	25.00–53.25			*
Lime, dolomite	45.00–67.00	*		

Radionic Fertilizer Rates (cont.)

Item	Rate	Source 1	Source 2	Source 3
Limestone	59.50–51.00	*		
Manganese sulfate	41.75–81.50			*
Manure, cow	61.75–51.75	*	*	*
Manure, horse	49.50–58.50	*	*	*
Manure, hog	68.50–58.75	*	*	*
Manure, poultry	56.50–37.20		*	*
Manure, poultry (BD composted)	30.75–25.00			*
Manure, rabbit	45.50–20.75	*		
Manure, sheep	10.00–36.00	*		
Molasses	22.80–27.50	*		
Molybdate, sodium	29.00–58.25			*
Mushroom waste	54.00–69.50		*	*
Nettle leaves	38.50–48.00		*	
Nitrate of soda	19.25–29.25	*		
Ntron (enzyme)	27.00–29.00			*
Oat hulls	32.00–24.50	*		
Peat, Irish	43.25–26.00		*	
Peat, general	27.00–37.80		*	
Perlite	43.00–43.25	*		
Phosphate, hard rock	42.50–48.00	*		*
Phosphate, soft rock (Cal-Phos)	29.25–53.00	*	*	*
Potassium-magnesium sulfate (Sul-Po-Mag, K-Mag)	41.00–59.75			*
Potassium sulfate	57.00–92.00		*	
Salt	82.00–37.00	*		
Sawdust	30.00–53.75	*		
Sea water	36.00–92.50		*	*
Seaweed	27.00–30.75			*
Seaweed extract	52.50–77.00	*	*	*
Sodium bicarbonate	21.00–45.00	*		*
Slag, basic	54.00–50.25		*	*
Sulfur, elemental	77.00–94.00	*	*	*

Balancing Individual Crops and Plants

Radionic Fertilizer Rates (cont.)				
Item	Rate	Source 1	Source 2	Source 3
Tea tree oil	55.25–30.00	*		
Vermiculite	55.25–82.00	*		
Vinegar	12.25–61.50	*	*	*
Vitamin C	25.00–29.50			*
Volcanic ash	65.50–54.50		*	*
Willard Water	12.50–37.75			*
Willard Water (black)	65.50–68.50			*
Wood ash	30.00–41.80		*	
Worm castings	52.50–61.00			*
Zinc oxide	32.50–66.50		*	*
Source 1: Lutie Larson, *Gardening Rate Book*, Pleasant Grove, UT: Little Farm Research, 1988. **Source 2**: Peter Kelly, *Psychotronics and Agriculture*, Lakemont, GA: Interdimensional Sciences, 1983. **Source 3**: T. Galen Hieronymus, *Cosmiculture*, Lakemont, GA: ASR&D, c.1980s.				

Research and development on cell-salt fertilizers continues. Researcher and entrepreneur Lee Roberts has developed a commercial product called SmartNute®.[9] He provided me with an early iteration of it, which I've applied directly and also broadcast using the rate 73.00–19.50. It proved more than satisfactory, but I understand Lee has already developed further enhanced versions.

I am also optimistic about several products that the Perelandra Center for Nature Research markets. *Essence of Perelandra* or *EoP* is a potency created to provide "additional balance, stability, strength and support" to the person, animal, or landscape to which it is applied.[10] Perelandra also markets a series of potencies labeled *ETS*, which are intended to relieve stress. One of these products is specifically formulated for plants.

---

9   See https://www.smartnute.com.

10 "Essence of Perelandra (EOP)," Perelandra Center for Nature Research (https://www.perelandra-ltd.com/Essence-of-Perelandra-EoP-P3195.aspx).

What are Cell Salts?

Cell salts are homeopathic preparations of certain mineral compounds, which are found within all living things.[11]

The cell salts, also known as *tissue salts*, were developed by the German doctor and homeopath Wilhelm Heinrich Schuessler (1821–1898). He determined that there are twelve mineral compounds important to the proper functioning of cells. A table of the cell salts (now thirteen in number) and their radionic rates follows:

Cell Salts and Their Radionic Rates		
Cell Salt	*English Chemical Name*	*Radionic Rate**
Calc. fluor.	Calcium fluoride	24.00–04.00/85.00–72.00
Calc. phos.	Calcium phosphate	92.00–96.00/24.00–04.00
Calc. sulph.	Calcium sulfate	24.00–04.00/77.00–94.00
Ferr. phos.	Ferric phosphate	92.00–96.00/49.00–27.00
Kali. mur.	Potassium chloride	30.50–67.00/37.00–93.00
Kali. phos.	Potassium phosphate	30.50–67.00/92.00–96.00
Kali. sulph.	Potassium sulfate	30.50–67.00/77.00–94.00
Mag. phos.	Magnesium phosphate	27.00–13.00/92.00–96.00
Nat. mur.	Sodium chloride	82.00–42.00/37.00–93.00
Nat. phos.	Sodium phosphate	82.00–42.00/92.00–96.00
Nat. sulph.	Sodium sulfate	82.00–42.00/77.00–94.00
Silica	Silicic oxide	89.50–91.50
LiBe	Lithium beryllium	11.00–42.00/69.75–79.00

* Most of the rates shown are called *two-bank rates* and were originally intended for instruments with two dial banks. If using a single-bank or other psychotronic broadcaster, make an intent card (source rates: Lutie Larsen, *The Experimental Ratebook*, Pleasant Grove, UT: Little Farm Research, 1988, pp. 39–41).

Controlling Diseases and Insect Pests?

As organic systems mature, there is a natural decline in pressure from many pest insects, mites, and diseases. I see that scenario beginning to play out in my garden. Biodynamic practices stiffen the spine of crops and soils and make them more resilient. Still, infestations and infections do show up now and then.

---

11 Jayne Marquis, "Cell Salts: How They Work" (https://www.betterhealthclinic.ca/bhcblogs/cell-salts-how-they-work).

Balancing Individual Crops and Plants

When that happens, doing nothing is sometimes the best response. Nature might already be in the process of restoring balance in ways we don't easily recognize. Several years ago, I visited some market gardeners who said that small Gila monsters were attacking their crops. They had painstakingly hand-picked and crushed all they could find, but more kept showing up. As most organic growers know, the larval stage of ladybird beetles looks remarkably like a tiny Gila monster. Both ladybird adults and their larvae are beneficial insects that eat aphids—the real pests that were attacking the plants.

Sometimes, when plants are diseased or heavily infested, the most practical action is to destroy them and prevent an isolated problem from spreading. There are other organically recommended means of prevention and control, such as choosing resistant varieties, hand-picking pests, exclusion by using row covers, companion planting, and releasing beneficial insects. Those practicing psychotronics have an added tool. We can use broadcasting to help suppress both diseases and pests. The table on the next page features a partial list of useful rates for identifying and controlling diseases, insects, and mites in tomatoes.

Poisons should always be a last-choice option. Even organically acceptable pesticides have health and environmental impacts well beyond killing a targeted pest. That doesn't mean you cannot or should not use them if your dowsing so indicates. Nature might decide it is the best option under your current circumstances. I like the term *rescue chemistry*, which I first read in an early *Acres U.S.A.* magazine.[12] It appropriately names agricultural pesticides and allopathic medicines when we use them to get us out of a bind, whether to save a crop or our life. At the same time, we should not make cornerstones of either our farming or gardening system or our health plan. Amen to that!

---

12 Acres U.S.A., P.O. Box 1690, Greeley, CO 80632 (www.acresusa.com).

Psychotronics and a Biodynamic Garden

Tomato Diseases and Insect Pests with Radionic Rates			
Disease or Condition	Rate	Pest Insect or Mite	Rate
Anthracnose	04.50–42.75	Aphids (1)	32.25–47.50
Asters yellows	33.50–85.50	Aphids (2)	13.00–59.00
Bacterial speck	28.50–79.75	Aphids (3)	35.75–57.00
Bacterial spot	31.75–42.50	Aphids (4)	38.25–19.25
Blight, early	32.50–21.00	Aphid, winged adult	33.75–67.75
Blight, late	26.50–55.50	Aphid, wingless adult	36.50–54.25
Blossom end rot	29.00–56.00	Aphid eggs	62.00–75.40
Gray leaf mold	24.75–88.25	Aphid nymphs	52.60–65.00
Gray leaf spot	35.50–62.00	Flea beetles (1)	41.50–71.25
Mildew, downy	58.50–67.50	Flea beetles (2)	43.00–67.00
Mildew, powdery	55.00–52.00	Flea beetles (3)	35.75–57.00
Pythium fruit rot	38.50–56.50	Flea beetles (4)	54.50–15.50
Septoria leaf spot	44.75–28.50	Flea beetle larvae	69.75–28.50
Tobacco mosaic virus (1)	23.00–08.00	Fruitworm, tomato (1)	33.00–70.00
Tobacco mosaic virus (2)	23.50–13.00	Hornworm, tomato	23.50–29.00
		Japanese beetle (1)	43.00–67.00
		Japanese beetle (2)	32.00–12.75
		Mites, red spider	38.75–24.75
		Stinkbugs	41.50–71.75
Rate sources include Peter Kelly, *Psychotronics and Agriculture*, Lakemont GA: Interdimensional Sciences, 1983; Lutie Larsen, Gardening Ratebook, Pleasant Grove, UT: Little Farm Research, 1988; Gene Litwiler, handout materials for a Farm-in-Balance Workshop, Kansas City, MO, Nov. 1987.			

Appendices

Appendix 1

The Evolution of Radionics and Psychotronics for Farming and Gardening

Part 1

This is the first in a planned series of articles to be published in the USPA *Journal and Newsletter,* summarizing the history and evolution of radionics–psychotronics as applied to horticulture and the growing of field crops.[1] To those who know and study radionics as a modality for human health and wellbeing, applications to agriculture may seem ancillary. It's true that society sets a priority on our physical and mental health by supporting a gargantuan and costly medical system and infrastructure. Sadly, however, it often focuses more on treating symptoms than on correcting (or at least admitting to) the underlying causes of disease, which include devitalized and residue-laden food, as well as environments polluted by pesticides and other contaminants from modern farming and other sources. The food we eat; the water we drink; the air we breathe; are keys to health, and the way we farm and garden has so very much to do with the quality of all three... and more. The ongoing expansion of organic and local markets indicate a growing proportion of the public recognizes this as well, and they are putting their food-dollars on the line.

---

1   This article was originally published in three parts in the *USPA Journal and Newsletter of the United States Psychotronic Association:* vol. 2, no. 9, (Sept. 2016) pp. 20–23; vol. 3, no. 1, (Jan. 2017) pp. 33–37, and vol. 3, no. 2, (Feb. 2017), pp. 39–45.

Radionics–Psychotronics can contribute to improvements in our food system and the wider environment in numerous ways. Two of these are obvious. First, we can use their analytical methods to assess the purity and vitality of foods in stores, farmers' markets, and on our plates. This application is already a frequent part of basic education in dowsing and radionics.

Second, radionics and psychotronics can be used directly in the growing of food, feed, fiber, and energy crops to increase production; suppress weeds, pests, and diseases; cut costs; reduce the pollution and contamination of soil, air, water, and food; and enhance discrimination among crops and crop varieties often developed through misguided breeding programs and objectives. By extension, we can use these radionics and dowsing modalities to assess the quality of foodstuffs arising from different systems of growing, plant breeding, and handling. This feedback can guide advancements in farming and gardening in exceptional ways.

UKACO and the Emergence of the Homeopathic Hypothesis

The story of modern agricultural radionics begins with the commercial enterprise called UKACO, and its spin-offs—the Homeotronic Foundation and the Radiurgic Corporation, which operated from the late 1940s through the early 1970s.[2] The name, UKACO, derives from the first letters of the last names of Curtis Upton, William Knuth, and Howard Armstrong—the principals of the company. Of these, Upton stands out as the main force behind the

---

2 UKACO's work began in the late 1940s. Precisely how long the entity continued to function is unclear. In 1952, the principals, particularly Upton, were instrumental in creating the Homeotronic Foundation—a not-for-profit organization whose purpose was the pursuit of the science behind agricultural radionics. The Radiurgic Corporation was another commercial venture that, in turn, emerged from the Homeotronic Foundation to again offer radionics services to growers. According to E. W. Russell, the Homeotronic Foundation was still active in 1973 when his book *Report on Radionics* was published.

Appendix 1

enterprise. Edward W. Russell, the author of *Report on Radionics*,[3] which is the primary source of information on UKACO, considered Upton the discoverer of agricultural radionics, much as he recognized Albert Abrams as discoverer of radionics.

UKACO is best remembered for its numerous successes in radionic pest control—most of which were on a large scale—on crop fields in Pennsylvania and Arizona. They employed remote broadcasting, using aerial photographs as witnesses and botanical pest control substances[4] as reagents to rid crops of pest insects. The operators delimited the treated areas by drawing boundaries on the photos and/or cutting away non-target zones. This provided for "check plots," which could be compared to the treatment to assess the effectiveness of the broadcasts. Most of these treatments appear to have been very successful; numerous field results are summarized in the aforementioned *Report on Radionics*. Also included are the verbatim texts of letters of documentation and support from the Farm Bureau, which observed the field demonstrations. UKACO charged the growers on a per-acre basis only when treatments were successful. The cost was well below that of conventional pesticides and was of significant savings to farmers. There were many growers who were pleased with and supportive of this work.

The retelling of UKACO's pest control success stories has created a narrowed perspective and obscured what may be the true nature of radionics broadcasting to field crops and how it might actually work. The mythologized image suggests that pests were attacked or "zapped," at a distance, using souped-up radionic instruments and methods. This is a seductive image for a world immersed in drones, smart bombs, and violent video games. But while this image *might* be

---

3 Russell, *Report on Radionics: Science of the Future.*

4 The botanicals included pyrethrum, rotenone, nicotine, hemlock, horsetail, sumac, and dark red geranium. These materials were found to be at least as effective (if not more effective) than arsenic and mercury compounds, which were used as agricultural pesticides at that time. Arsenic- and mercury-based pesticides are now banned for most applications.

substantially accurate, there is a less sexy but more plausible explanation for radionic pest suppression—one built on concepts common to homeopathy, biodynamic farming, and organic agriculture.

To begin, many biodynamic and organic farming practitioners consider agricultural insect pests to have a specific role in nature as garbage collectors. They perceive their main function as the removal of plants that are sick, genetically damaged, or otherwise stressed and weakened. While these growers cannot control all stress factors, they understand many of them to be *agricologenic*—caused by the way we farm. Imbalanced crop nutrition, reaction to herbicides and other pestcides, destruction of soil tilth, and loss of biodiversity are among problems resulting in modern farming and gardening. Traditional biodynamic and organic growers, therefore, look first at what they are doing in the field which might have encouraged pests to get out of control, and then correct that, if possible.

This line of thinking is well supported by the overlapping theories of *predisposition*[5] and *trophobiosis*.[6] In a compelling example, it was discovered that stressed plants cease building proteins and instead begins to reduce them. This results in the proliferation of excess free nitrogen and amino acids in plant cells and sap. These compounds are the preferred foods of herbaceous insect pests, like aphids, which lack the enzymes to break down whole proteins into their amino acid constituents.[7] Assuming this to be true, insect

---

5   Eliot W. Coleman and Richard L. Ridgeway, 1983. "Role of Stress Tolerance in Integrated Pest Management," p. 126, in Knorr, *Sustainable Food Systems*.

6   "The theory of trophobiosis has to do with how plant nutrition affects plant health, with what makes a plant susceptible or resistant to disease and to pest attack" (J. A. Lutzenberger), in Chaboussou, *Healthy Crops: A New Agricultural Revolution Charlbury*.

7   Eliot Coleman, as quoted by Stoner, Kim, and Tracy LaProvidenza. 1998. "A history of the idea that healthy plants are resistant to pests," p. 4, in K. Stoner, "Alternatives to Insecticides for Managing Vegetable Insects: Proceedings of a Farmer/Scientist Conference (NRAES-138)," NRAES, Ithaca, NY, Dec. 6–7, 1998; also. "Pests starve on healthy plants," *Ecology Action Newsletter*, Willits, CA, May 1999, pp. 3–4.

Appendix 1

pests on healthy plants—being denied the digestible food that sick or stressed plants provide—need not be directly attacked either with pesticides or radionics; they will either starve in place or seek nourishment elsewhere.

This might help explain the observations by radionics pioneer Peter Kelly, as expressed in an interview from the early 1980s.[8] In response to a question about the corporeal remains of insect pests controlled by radionic broadcasting, Peter notes that simple organisms, such as corn borers, essentially dissolve into "native materials like water and basic energy"; more complex organisms, on the other hand, "would have some remnants." This almost certainly reflects what one would likely observe under circumstances where a crop suddenly becomes unpalatable to pest insects. The larval or caterpillar (primitive) stage of moth, butterfly, beetle, and fly pests—being soft-bodied and unable to migrate to other feeding grounds, are likely to die of starvation in place and decompose rapidly. Adult forms and those insect pests that do not pass a larval stage (e.g., true bugs) have an exoskeleton that would resist immediate degradation and leave some remains. However, such insects are also more mobile and better equipped to move on!

Peter Tompkins and Christopher Bird hint at an alternative to the "zapping" hypothesis in their 1973 book *The Secret Life of Plants*.[9] They suggest that UKACO's radionic broadcasts may have acted homeopathically—directly affecting the plants (as opposed to the pests)—stimulating natural resistance. If true, we might argue that this resistance may have assumed a plant-nutritional form to counter insect feeding, as suggested by the predisposition and trophobiosis theories.

This notion takes on more weight when one carefully reads Russell and Tompkins and Bird. Both sources indicate that UKACO sought

---

8 Leslie Aickin, interview, in "Psychotronic Farming," pp. 8–13, in *Kelly, Psychotronics Book 1*.

9 Tompkins and Bird, *The Secret Life of Plants*, p. 323.

not merely to eliminate pests, but also to vitalize crops to increase yield and quality! If we assume that successful vitalization equates to stress reduction, which further leads to pest starvation, it is logical that most observers might measure success in terms of pest suppression. They might simply note any improvement of crop performance as resulting only from reduced pest damage. It would be quite possible to overlook the underlying causal factors associated with healthier plants.

It is not clear from Russell and Tompkins and Bird, as well as other published sources exactly what protocols UKACO principals and cooperators followed, and whether the alternative "homeopathic–predisposition–trophobiotic" explanation truly holds water. Nevertheless, we might take a hint from another radionics pioneer, T. Galen Hieronymus, who worked with UKACO's not-for-profit research arm, the Homeotronic Foundation.

In the *"credimus"*[10] or introduction to his *Cosmiculture* manual, Galen outlines his philosophy and approach to using radionics–psychotronics for agriculture. He states:

> Disease, unwanted insects, undesirable plants are simply indications of conditions, in that environment, conducive to their existence at a particular time and place. Change those conditions by enhancing the environment for the desirable, and the reason for the undesirable ceases to exist. Reagents are incorporated within the Cosmiculture system that will, at once, enhance the vitality of the desirable *and* reduce the vitality of the undesirable.[11]

This statement of belief mirrors the ideas common to predisposition and trophobiosis, as well as biodynamic, homeopathic, and organic farming traditions. They might also have been those of Upton, Knuth, and Armstrong. Regardless, the concept continues through the contemporary evolution of agricultural radionics, the story of which will continue in Part 2 of this series.

---

10   Roughly translated from the Latin, *credimus* means "what we believe."
11   Hieronymus, *Cosmiculture*, p. 1.

Appendix 1

PART 2

***Part 1** briefly reviewed the history of UKACO and its pioneering role in agricultural radionics, focusing on its successes in large-scale pest control using remote broadcasting, and discussing alternative interpretations of the mechanisms involved. **Part 2** continues with contributions from George and Marjorie de la Warr to radionics for horticulture.*

George and Marjorie de la Warr

English practitioners George "Bill" and Marjorie de la Warr were pioneers of radionics and advanced the original work of the American originator, Albert Abrams, his "successor" Ruth Drown, and others. The de la Warrs were contemporaries of Curtis Upton and Upton's UKACO collaborators, developing radionics instrumentation, protocols, and theories primarily during the 1940s through the early 1970s. They were truly the first to produce highly accurate standardized radionics instruments.[12]

While they worked as a team, the couple divided their labors: Marjorie focused on developing a successful radionics enterprise, treating patients; George was involved primarily in research and development. Thus, most of the inventions, scientific findings, and theories discussed in the literature are usually credited to him.

The de la Warrs not only built on others' ideas but brought highly original ideas and innovations to the field of radionics. George is noted for, among other things, illuminating concepts of resonance and force-field bodies and how fundamental energy ultimately manifests as matter.[13]

The de la Warrs are seldom mentioned when speaking about agricultural radionics. In fact, George did much of his research

---

12 Russell, *Report on Radionics: Science of the Future.*
13 Laurie, *The Secret Art: A Brief History of Radionic Technology for the Creative Individual,* pp. 83–99.

on plants, which he preferred as research subjects. However, most of the study was done on a noncommercial, horticultural-garden scale; thus, the tendency of the agricultural radionics world to overlook it. That said, the discoveries made were not merely interesting, but significant to our understanding of how radionics works in all its applications. It is certainly worthy of our attention.

Horticultural Research

One of the first things this writer noticed when reading accounts of George's research was the attention to scientific protocols—essential for earning the serious attentions of serious scientists. In agronomic and horticultural research, one of the most important problems to overcome is soil variation. Differences in texture, structure, organic content, and minerals are often quite significant over short distances on what appear to be uniform fields and gardens. In one notable instance, George de la Warr overcame soil variation by physically removing the soil from the test plots and sieving and mixing it together before returning it to the experimental beds. Though the trial lacked replication, this exacting procedure demonstrated his determination to use good scientific protocols to generate unbiased and valid findings. This particular trial evaluated the effect of radionically treated soil on transplanted cabbage. When compared, the treated plot produced plants three times as large as those from the control plot.[14]

In a follow-up trial—done later on a site with naturally more uniform soil—he employed a replicated and randomized plot layout. This is the more common approach used by agronomists and other scientists to overcome site variations. In this trial, the de la Warrs substituted broccoli for cabbage. At termination of the trial, plants from treated plots averaged eighty-one percent more weight than those from the untreated control plots. Furthermore, plants from

---

14 Day (with George de la Warr), *New Worlds beyond the Atom*, pp. 79–80.

treated plots revealed no necrosis, while this was evident in the control plants.[15]

One of the few larger-scale trials entailed treating a commercial tobacco crop in Rhodesia. Radionic treatments caused early flowering—an undesirable occurrence in a crop where leaves are the desired product. Still, the leaf quality proved exceptional and the crop fully marketable.[16]

In another set of experiments, George germinated plants in radionically treated and untreated vermiculite, noting improved growth where radionic treatments were applied. Though de la Warr's results were consistent when done in his lab, they failed when independent researchers tried to duplicate his work! George concluded, after this and other similar failures, that the plants in these trials were not responding to radiations directly from instrumentation, but indirectly to those originating from the humans operating the instruments.[17] This was a highly important observation, with implications well understood by the radionics community today. The influence of the human element in radionics continues to bedevil efforts to gain credibility for our science and practice.

Critical Rotation Position

One of the most heralded discoveries by de la Warr, vis-à-vis plants, relates to his discovery of the *critical rotation position* (CRP), which refers to the optimal rotational positioning of plants relative to the earth's magnetic field. A plant in CRP is in "optimal resonance with the life sources that sustain it;...it is receiving the optimum quantity of vital radiations."[18]

The CRP matters particularly for potted plants since they are frequently rotated to encourage uniform growth. Crops, fruit trees,

---

15 Ibid., p. 80.
16 Ibid., p. 81–82.
17 Tompkins and Bird, *The Secret Life of Plants,* pp. 347–348.
18 Day, *Op. cit.,* pp. 80–81.

and other plants that are transplanted are likely to suffer some positional stress as well, since few growers are even aware of the CRP concept. Direct-seeded plants and crops are another matter. We understand that seeds planted directly into the soil naturally orient themselves to their CRP as they germinate and grow.[19] Obviously, the same natural orientation occurs with plants started in pots or trays. It is our subsequent repositioning of them that leads to loss of CRP resulting in stress to the plants.

One can determine the CRP through dowsing while physically rotating the potted plant or the transplant before setting it.[20] When growing field transplants, this writer marks the north-facing aspect of seedling trays and maintains that orientation throughout greenhouse growing, hardening, and final placement in the field. This is doable in hand-planting and probably impossible in mechanized transplanting.

Perhaps we should not describe de la Warr's discovery of CRP as *discovery*, but as rediscovery. Dr. Albert Abrams, the father of radionics, made the initial observations that ultimately led to radionics while performing a standard analytical technique called percussion. While percussing the abdomen of a patient, he detected some anomalies that he chose not to ignore. After percussing the patient in every position possible, he noted that the response he found was strongest when the patient faced west. Such phenomena were observed with other patients, demonstrating that humans also have something like critical rotation position.[21]

Finding the Front Door

The concept of critical rotation position for plants harkens back to the notion of plants having *energy doors*—specifically, a *front* door. This is a familiar concept to many dowsers. Among its

---

19 Moore, *Stone Age Farming: Eco-agriculture for the 21st Century*, p. 71.
20 Ibid., p. 71.
21 Russell, *Op. cit.*, pp. 15–17.

practical uses is in the proper placement of French or Lakhovsky coils around tree trunks. Made of copper wire, these coils are used for healing or to enhance tree growth.

Australian dowser Alanna Moore claims that the front door of a plant is part of its aura—a spot where a concentrated flow of energy can be detected, functioning perhaps as the heart chakra or the seat of consciousness for the plant.[22]

According to Harvey Lisle, "…the energy door of a tree is associated with the neutral point of a magnet—located where the positively charged atmosphere meets the negatively charged Earth."[23] Lisle's description appears to correspond with the location defined by Lutie Larsen for the *physical support center* or *chakra*—one of four major "chakras" she has identified for plants.[24]

A Radionic Camera

Another de la Warr innovation that truly stands out is their radionic camera. Like most radionic instruments, the de la Warr camera worked with witnesses such as blood spots to produce images of organs and the fetuses of pregnant women. When they used the camera to analyze seeds, the de la Warrs could, by adjusting rates, produce images of the plant or plant parts—such as flowers—that the seed had the potential to produce later on. In doing so, they reinforced the theory of *Life-* (or *L-*) *Fields* as advanced by Yale Professor of Anatomy Harold Saxton Burr[25] and outlined in his writings, including his popular book *The Fields of Life: Our Link with the Universe*.[26]

Essentially, Burr asserts that all living things have *electrodynamic fields* that can be detected and measured using standard

---

22 Moore, *Op. cit.*, pp. 71–72.
23 Ibid., p. 181.
24 Larsen, *Little Farm Tips and Techniques for Farmers*, pp. 134–135.
25 Laurie, *Op. cit.*, p. 96.
26 Burr, *The Fields of Life: Our Link with the Universe*.

Psychotronics and a Biodynamic Garden

voltmeters, and that these fields are the blueprint and control for each organism's development, health, and mood.

The de la Warrs were unaware of Burr's theories at the time, yet they reached very similar conclusions and came to the firm belief that radionics works with the pre-physical, in contrast to the momentary, physical body or form.[27] This is a principle most practitioners hold to be true today.

Lutie Larsen does a good job of elaborating on this concept when writing about *archetypal patterning,* which she describes as "an active energetic recall between the plant and its original formative template."[28] Whereas the ideations of Burr, de la Warr, and Larsen are probably not identical, the notion that subtle information fields exist as templates guiding the growth and development of living organisms—within their archetypal frameworks—is certainly held in common.

---

27 Russell, Edward W., *Op. cit.* p. 157.
28 Larsen, *Op. cit.*, p. 114.

Appendix 1

Part 3

Parts 1 & 2 of this series focused on the contributions of UKACO and George and Marjorie de la Warr to radionics for agriculture and horticulture. The third and final part addresses the work of T. Galen Hieronymus, Peter Kelly, Hugh Lovel, and other contemporaries.

T. Galen Hieronymus

Thomas Galen Hieronymus (1895–1988) was one of the truly great pioneers of radionics. His long life and career made him a contemporary of everyone in the field for about three-fourths of the twentieth century. Galen's investigations addressed basic research into the nature of the energy (or energies) with which he and others were working. He coined the term *eloptic* to describe this energy. The word *eloptic* is derived from *electricity* and *optics*. Hieronymus observed that the energy exhibited some behaviors associated with electricity, and some associated with light.[29] In the process of researching, he designed and built instrumentation that set the pattern for many (perhaps most) of the radionic and psychotronic instruments built and used in the United States today.

As mentioned in Part 1 of this series,[30] T. Galen Hieronymus worked with UKACO's not-for-profit research arm, the Homeotronic Foundation. Galen's autobiography,[31] published posthumously in 1988, details his efforts to address various plant diseases and insect pests. Among these were apple scab, Dutch elm disease, tomato early blight, tomato hornworm, corn earworm, tent caterpillar, spruce budworm, aphids, citrus mealy bug, and nematodes.[32]

---

29 Sarah Hieronymus (ed.), *The Story of Eloptic Energy: The Autobiography of an Advanced Scientist Dr. T. Galen Hieronymus*, p. 213.

30 Kuepper, "The Evolution of Radionics and Psychotronics for Farming and Gardening, Part 1," *USPA Newsletter*, vol. 2, no. 9. Sept. 2016. p. 23.

31 Hieronymus, Sarah (ed.), *The Story of Eloptic Energy*.

32 Ibid., pp. 282–292, 447.

Psychotronics and a Biodynamic Garden

Whereas Hieronymus, his team, and various cooperators conducted other kinds of work with plants and crops, earlier work focused on isolating, and then killing and/or repelling such pests and diseases. In this regard, it is important to note that Galen anticipated that this approach could have collateral consequences to crop plants. He stressed that the reagents selected for radionic pest control must *not* be harmful to the plant or crops, no matter their effect on the target organism.[33]

This becomes evident when studying his protocols for eloptic pest control. The following is a verbatim description of these protocols as included in his autobiography:

> First, we isolate a specimen of the pest. If it be a larvae, we find one and put it in a test tube. Next, we take a leaf from the host plant and put it in another test tube. Then with the test tube containing the pest in the [eloptic or radionics] analyzer, we measure his vitality. Next, we try first one then another of the reagents we think might be effective in lowering the vitality of the pest.
>
> When we have found such a reagent, we try it with the leaf from the plant to be sure we will not poison the plant while we destroy the pest. We prefer to use a reagent that will be good for the plant at the same time it destroys the pest. Usually we can find a reagent that works properly, however, sometimes it requires the use of two or more reagents....
>
> We broadcast the energy from the plant leaf and the reagent, but out of phase 180 degrees. It neutralizes the normal energy from the pest when it eats from the plant so irradiated.... It took years of work with many methods of procedure to arrive at this simple method. An analyzer has been developed that will tune in to the vibrations or emanations from a pest or a plant and measure its vitality.
>
> The reagents consist of a small amount of each material in a test tube. Some are poisons, some are herbs, some are plant

---

33 Ibid., p. 284.

oils, some are antibiotics, anything that gives promise of being effective....

Eloptic energy will travel on light. We can take a photo on black and white or polaroid film and use it as the specimen of the plant instead of a leaf from the plant. Aerial photos work very effectively. In this way, we can treat large areas at one time....

No pest seems to be immune if we can find a reagent that will react to lower its vitality. So far, we have been successful in finding the required reagent and the pests worked with have been many and varied.[34]

Isolated Kill vs. Trophobiosis

In his autobiography, Hieronymus elaborates further on radionic pest control, discussing the fate of targeted insects. He theorizes that death ultimately results from offsetting "binding energy," which allows the component elements to return to their original state, thus—in the case of larvae—leaving only "moist spots."[35] Peter Kelly (1948–2004)—another mid- to late-twentieth-century radionics pioneer—addressed this matter similarly in a 1984 interview. In describing the fate of radionically "killed" corn borers, he said that, being simple organisms, they reverted to native materials like water and basic energy—that they virtually disappeared.[36]

It may seem to be a fine point, but this "isolate and kill" approach, while certainly valid in its own right, largely ignores the possibilities of *predisposition* and *trophobiosis*, which suggest something like a homeopathic mode of action. These theories (predisposition and trophobiosis) propose that: 1) truly healthy plants are not seriously attacked by most insect pests; 2) stress factors cause physiological changes within plants that make them attractive to, and digestible by, herbivorous insect pests; and 3) many

---

34 Hieronymus, Sarah, *Op. cit.*, p 284.
35 Ibid., p. 139.
36 Leslie "An interview with Peter Kelly," p. 9 in Kelly, *Psychotronics, Book 1.*

stress factors are nutritional and *agricologenic*—caused by poor farming practices and inputs; 4) changing farming and gardening practices to those that restore healthy balance is the first step toward reducing or eliminating a pest problem.[37] These ideas are exemplified in the philosophy and practice of biodynamics,[38] where pest infestations are not seen as inevitable problems but as indications of imbalances that need to be corrected.

Hieronymus acknowledged this philosophy somewhat later in his work and incorporated it into eloptic farming and gardening, which he began calling *Cosmiculture*. In the *Credimus*,[39] or introduction to the Cosmiculture manual, Galen outlines his philosophy and approach to using radionics/psychotronics for agriculture:

> Disease, unwanted insects, undesirable plants are simply indications of conditions, in that environment, conducive to their existence at a particular time and place. Change those conditions by enhancing the environment for the desirable, and the reason for the undesirable ceases to exist.
>
> Reagents are incorporated within the Cosmiculture system that will, at once, enhance the vitality of the desirable <u>and</u> reduce the vitality of the undesirable.[40]

Peter Kelly also expressed his support for the ideas behind predisposition and trophobiosis. In the aforementioned interview, he goes on to reference the work of Phil Callahan[41] and states, "Rather

---

37 Kuepper, *Op. cit.*, pp. 21–22.

38 "Biodynamics is a holistic, ecological, and ethical approach to farming, gardening, food and nutrition. Biodynamic agriculture has been practiced for nearly a century, on every continent on Earth. Biodynamic principles and practices are based on the spiritual insights and practical suggestions of Dr. Rudolf Steiner, and have been developed through the collaboration of many farmers and researchers since the early 1920s" (description from the Biodynamics Assoc. Dec. 2016: See https://www.biodynamics.com/what-is-biodynamics.

39 Roughly translated from the Latin, *credimus* means "what we believe."

40 Hieronymus, *Cosmiculture*, p. 1.

41 Philip Callahan is a retired USDA entomologist noted for his energetic views on nature, most notably, perhaps, for his theory that insect pests are

than trying to kill anything (since everything has its place in nature somewhere, even the insects in cleaning up diseased or unbalanced crops), would be to raise the vitality of the plant, raise the vitality of the field, so that the insects are no longer attracted to them."[42]

Cosmic Pipes

In 1984, Hieronymus introduced a new type of psychotronic instrument at the Denver conference of the Western Psychotronics Association. It was intended for agricultural and environmental applications, and he called it the *cosmic pipe*.[43] Cosmic pipes are built from weather-resistant PVC pipe and intended for outdoor installation. Units are ten feet long, two feet of which is buried in the ground with eight feet exposed, like an upright tower. Cosmic pipes are designed to take in, amplify, and redistribute cosmic energy modulated by reagents selected to enhance the environment, stimulate crops, and/or control pests. Hugh Lovel—a Georgia farmer and cooperator of Galen, who, incidentally, introduced him to biodynamics—describes cosmic pipes as stationary, self-driven instruments that directly induce self-reinforcing, resonant, fractal patterns as homeopathic potencies into the life energy of fields—day and night, 365 days a year.[44]

Lovel is an acknowledged leader in agricultural radionics and has continued to work with cosmic pipes, renaming them *field broadcasters*—improving on their design, management protocols, and performance. Among several advancements are:

---

    drawn to unhealthy crops by infrared radiations. This topic is covered in his popular 1975 book, *Tuning in to Nature: Infrared Radiation and the Insect Communication System*.

42 Aickin, *Op. cit.*, pp. 9–10.

43 Hieronymus, Sarah, *Op. cit.*, p. 440.

44 Lovel, *Quantum Field Broadcasting Comes of Age*, informational booklet, p. 1.

Psychotronics and a Biodynamic Garden

Field–farm boundary establishment: Initially, cosmic pipes were installed with no limitations on the area to be treated. Neighboring lands would receive the same broadcasts as the target farm, usually to their benefit. However, these unrestrained broadcasts would diminish with distance from the unit. Lovel borrows a page from UKACO, using aerial photos and maps with targeted field and farm boundaries indicated. This results in enhanced and uniform broadcasting only within the targeted area.[45]

Treatment of the atmosphere: The Hieronymus design treated the soil system only, occasionally resulting in imbalances. With his background in biodynamics, Hugh recognized the value of treating not only the soil, but the atmosphere into which the aerial portions of the crop grow—where photosynthesis, blossoming, fruiting, ripening, and other processes occur.[46] Biodynamics distinguishes between the *calcium process,* which encompasses downward, or earthy, patterns, and the *silica process,* which encompasses upward, or cosmic, patterns.[47] To accomplish this Hugh redesigned the original cosmic pipe with a second reagent well and modifications that permit broadcasting distinctly different modulated energy patterns concurrently to the atmosphere and the soil.

Advances in biodynamics applied to field broadcasting: Hugh was among the first to promote the use of a new BD Preparation—horn clay—a remedy that the father of biodynamics, Rudolf Steiner, had not worked out before his death. Horn Clay stimulates the proper flow of plant sap.[48] Horn clay variations are made for fall to spring (downward sap flow) and spring to fall (upward sap flow).

---

45 Ibid., pp. 1–2.

46 Lovel, Hugh. No date. *Twenty-first Century Field Broadcasting,* informational booklet, pp. 3–4.

47 Lovel, *Quantum Agriculture: Biodynamics and Beyond,* p. 41.

48 Lovel, *Twenty-first Century Field Broadcasting,* pp. 4–5.

Appendix 1

Towers of Power

Around the same time that Hieronymus was developing and promoting his cosmic pipe, another important pioneer of the period, Jerry Fridenstine, was researching and installing a different kind of psychotronic tower system on farms. These were called Triune Bio-tronic Tower Balancers[49] or *Towers of Power*. Compared to cosmic pipes, the biotronic units were more skeletal in appearance, with three long legs arranged in a three-sided pyramid form and a tube for holding a selection of reagents. Installation required the ability to understand and identify earth energy lines and, especially the crossing points of those lines, where energy flows in and out of the earth. This information was essential for locating the towers. It is unclear from the Hieronymus literature, however, whether his cosmic pipes should be installed in a similar manner with consideration of earth grid lines.

Radionics and Farm Inputs

In the Aickin interview, Peter Kelly goes beyond pest control to briefly outline radionics methods for selecting and balancing seeds, and for selecting and potentizing crop fertilizers.[50] In so doing, he describes the approach taught by many radionics educators from the 1980s onward, including such notables as Steve Westin, Phil Wheeler, Arden Andersen, and Jerry Fridenstine. This approach emphasized using instruments to select the optimal fertilizers and soil amendments, "clean" them radionically, potentize them, and apply them physically to soils and crops. Please note, however, that these instructors were not, and are not, abandoning remote balancing or other traditional radionics methods. Rather, they are expanding the ways in which radionics can be used in farming and

---

49 Fridenstine, *What Are Triune Bio-tronic Tower Balancers?* informational literature.

50 Aickin, *Op. cit.*, pp. 10–11.

– 201 –

gardening by generally reducing the quantity of commercial agricultural inputs in an effort to improve plant growth, save money, and reduce environmental damage.

Summary

I wrote this series of articles as a brief overview of radionics as it has been applied to agriculture. I have tried to highlight the principal researchers and developers, the contributions they made, and the ideas they espoused and shared. There are many others whose names were mentioned only in passing, and some were not mentioned at all. My apologies...

For those wishing to pursue more details on radionics for farming and gardening, there are several resources I would suggest. The best how-to books available on agricultural radionics are Lutie Larsen's *Little Farm Tips and Techniques for Farmers*,[51] and my own *Plants, Soils, Earth Energy, and Radionics*.[52] Two further recommendations that discuss radionics and other metaphysical approaches to farming and gardening are *Secrets of the Soil*[53] and *Stone Age Farming*.[54]

---

51 Larsen, *Little Farm Tips and Techniques for Farmers*.
52 Kuepper, Plants, *Soils, Earth Energy, and Radionics*.
53 Tompkins and Bird, *Secrets of the Soil*.
54 Moore, *Stone Age Farming: Eco-agriculture for the 21st Century*.

Appendix 2

Resources

Acres U.S.A.
P.O. Box 1690
Greeley, CO 80632
Phone: 970-392-4464
Email: info@acresusa.com
www.acresusa.com

American Society of Dowsers (ASD)
PO Box 24
Danville, VT 05828
Phone: 802-684-3417
Email: info@dowsers.org
www.dowsers.org

Biodynamic Association of North America
PO Box 557
East Troy, WI 53120
Phone: 262-649-9212
Email: info@biodynamics.com
www.biodynamics.com

Josephine Porter Institute (JPI)
652 Thompson Rd SE
Floyd, VA 24091
Phone: 540-745-7030
Email: info@jpibiodynamics.org
www.jpibiodynamics.org

United States Psychotronics Association (USPA)
525 Juanita Vista
Crystal Lake, IL 60014
Phone: 815-355-8030
Email: contact@psychotronics.org
www.psychotronics.org

Cited Works

Allen, Phil, et al. (eds.). *Energy, Matter, and Form: Toward a Science of Consciousness*. Boulder Creek, CA: University of the Trees, 1975.

Baan, Bastiaan. *Lord of the Elements: Interweaving Christianity and Nature*. Edinburgh: Floris Books, 2006.

Bird, Christopher. *The Divining Hand: The 500-Year-Old Mystery of Dowsing*. New York: Dutton, 1979.

Bortoft, Henri. *Taking Appearance Seriously: The Dynamic Way of Seeing in Goethe and European Thought*. Edinburgh: Floris Books, 2013.

———. *The Wholeness of Nature: Goethe's Way toward a Science of Conscious Participation in Nature*. Hudson, NY: Lindisfarne Books, 1996.

Burr, Harold Saxton. *The Fields of Life: Our Link with the Universe*. New York: Ballantine, 1973.

Callahan, Philip S. *Tuning in to Nature: Infrared Radiation and the Insect Communication System*. Austin, TX: Acres U.S.A., 1975 (rev. ed. 2001).

Carvallo, M.E. *Nature, Cognition and System II: Current Systems-Scientific Research on Natural and Cognitive Systems Volume 2: On Complementarity and Beyond*. Baden-Baden, Germany: Springer Science and Business Media. 2013

Chaboussou, Francis. *Healthy Crops: A New Agricultural Revolution*. Charlbury, UK: Jon Carpenter, 2007.

Cooper, Irving S. *Theosophy Simplified*. Wheaton, IL: Quest, 1979 (1st ed. 1915).

Copen, Bruce. *Electronic Homoeopathy for Plants*. Sussex, UK: Academic Publications, 1980.

Crombie, R. Ogilvie. *Encounters with Nature Spirits: Co-creating with the Elemental Kingdom*. Rochester, New York: Findhorn, 2018.

Day, Langston (with George de la Warr). *New Worlds beyond the Atom*. London: Vincent Stuart, 1956.

Engelken, Ralph, and Rita Engelken. *The Art of Natural Farming and Gardening*. Greeley, IA: Barrington Hall, 1981.

Erbe, Hugo. *Hugo Erbe's New Bio-dynamic Preparations*. Gloucestershire, UK: Mark Moodie, 2003.

Fridenstine, Jerry. *What Are Triune Bio-tronic Tower Balancers?* Reedsville, OH: Energy Refractors, n.d.

Goodavage, Joseph F. *Magic: Science of the Future.* New York: New American Library, 1976.

Hainsworth, P. H. *Agriculture: The Only Right Approach.* Pauma Valley, CA: Bargyla and Gylver Rateaver, 1976 (1st ed., Faber and Faber, 1954).

Hartman, Jane E. *Shamanism for the New Age: A Guide to Radionics and Radiesthesia.* Placitas, NM: Aquarian Systems, 1987.

Herbert, Nick. *Quantum Reality: Beyond the New Physics.* New York: Anchor Doubleday, 1985.

Hieronymus, Sarah (ed.). *The Story of Eloptic Energy: The Autobiography of an Advanced Scientist Dr. T. Galen Hieronymus.* Lakemont, GA: Institute of Advanced Sciences, 1988.

Hieronymus, T. Galen. *Cosmiculture.* Lakemont, GA: Advanced Sciences Research and Develipment (A.S.R. & D.), ca. 1980s.

Joye, Shelli Renée. *Developing Supersensible Perception: Knowledge of the Higher Worlds through Entheogens, Prayer, and Nondual Awareness.* Rochester, VT: Inner Traditions, 2019.

Kelly, Peter. *Psychotronics, Book 1* (3rd rev.). Lakemont, GA: Interdimensional Sciences, 1986.

Knorr, Dietrich (ed.). *Sustainable Food Systems.* Westport, CT: AVI, 1983.

Koepf, Herbert H. *Biodynamic Sprays.* Kimberton, PA: Biodynamic Farming and Gardening Association, 1988.

———. *Koepf's Practical Biodynamics: Soil, Compost, Sprays and Food Quality.* Edinburgh: Floris Books, UK, 2012.

———. *What is Bio-Dynamic Agriculture.* Wyoming, RI: Bio-Dynamic Literature, 1976.

Kuepper, George. *Heirloom Vegetables, Genetic Diversity, and the Pursuit of Food Security.* Poteau, OK: Kerr Center for Sustainable Agriculture, 2008.
———. *Market Farming with Rotations and Cover Crops: An Organic Bio-Extensive System.* Poteau, OK: Kerr Center for Sustainable Agriculture, 2015.
———. *Organic Bio-extensive Management Revisited.* Poteau, OK: Kerr Center for Sustainable Agriculture, 2017.
———. *Plants, Soils, Earth Energy, and Radionics.* Goshen, AR: GAIA, 1998.

Larsen, Lutie. *Little Farm Tips and Techniques for Farmers.* Pleasant Grove, UT: Wise Woman Venture, 2008.

Cited Works

Laurie, Duncan. *The Secret Art: A Brief History of Radionic Technology for the Creative Individual.* New York: Anomalist Books, 2009.

Leftwich, Robert H. *Dowsing: The Ancient Art of Rhabdomancy.* New York: Weiser, 1976.

Lisle, Harvey. *The Enlivened Rock Powders.* Greeley, CO: Acres U.S.A, 1994.

Lovel, Hugh. *A Biodynamic Farm: For Growing Wholesome Food.* Greeley, CO: Acres U.S.A., 1994, 2000.

———. *A Cosmic Pipe Update…Farming the Atmosphere.* Greeley, CO: Acres U.S.A., 1997.

———. *Field Broadcasting 25 Years On.* Greeley, CO: Acres U.S.A., 2012.

———. *Quantum Agriculture: Biodynamics and Beyond.* Blairsville, GA: Quantum Agriculture, 2014.

———. *Quantum Field Broadcasting Comes of Age.* Blairsville, GA: Union Agricultural Institute, n.d.

———. *Stimulating Soil and Air.* Greeley, CO: Acres U.S.A., 2002.

———. *Ten Years with a Cosmic Pipe.* Greeley, CO: Acres U.S.A., 1996.

———. *Twenty-first Century Field Broadcasting.* Blairsville, GA: Union Agricultural Institute, n.d.

Massey, Robert E. *Alive to the Universe: A Layman's Handbook of Supersensonics.* Boulder Creek, CA: University of the Trees, 1976.

McKanan, Dan. *Eco-alchemy: Anthroposophy and the History and Future of Environmentalism.* Oakland: University of California, 2018.

Moore, Alanna. *Divining Earth Spirit: An Exploration of Global and Australasian Geomancy.* Castlemaine, Australia: Python Press, 2004.

———. *Stone Age Farming: Eco-agriculture for the 21st Century.* Victoria, Australia: Python, 2001.

New Zealand Biodynamic Farming and Gardening Association. *Biodynamics: New Directions for Farming and Gardening in New Zealand.* Aukland, NZ: Random Century, 1989.

Paulson, Genevieve L. *Kundalini and the Chakras: Evolution in this Lifetime.* St. Paul, MN: Llewelyn, 1991.

Pogačnik, Marko. *Nature Spirits and Elemental Beings: Working with the Intelligence in Nature.* Forres, Scotland: Findhorn, 1995.

Russell, Edward Wriothesley. *Report on Radionics: Science of the Future*. Suffolk, UK: Neville Spearman, 1973.

Steiner, Rudolf. *Agriculture: Spiritual Foundations for the Renewal of Agriculture*. Kimberton, PA: Biodynamic Association, 1993.

———. *How to Know Higher Worlds: A Modern Path of Initiation*. Hudson, NY: Anthroposophic Press, 1994.

———. *An Outline of Esoteric Science*. Hudson, NY: Anthroposophic Press, 1997.

———. *What Is Biodynamics? A Way to Heal and Revitalize the Earth*. Great Barrington, MA: SteinerBooks, 2005.

Stine, G. Harry. *Mind Machines You Can Build: Move Things with Your Mind and Other Experiments*. Largo, FL: Top of the Mountain, 1992.

Stoner, K. *Alternatives to Insecticides for Managing Vegetable Insects: Proceedings of a Farmer/Scientist Conference* (NRAES-138). Ithaca, NY: NRAES, 1998.

Storl, Wolf D. *Culture and Horticulture*. Wyoming, RI: Bio-Dynamic Literature, 1979.

Stout, Ruth. *How to Have a Green Thumb without an Aching Back: A New Method of Mulch Gardening*. New York: Exposition Press, 1955.

Swann, Ingo. *Everybody's Guide to Natural ESP: Unlocking the Extrasensory Power of Your Mind*. Los Angeles: Tarcher, 1991.

Talbot, Michael. *The Holographic Universe: The Revolutionary Theory of Reality*. New York: HarperCollins, 1991.

———. *Mysticism and the New Physics*. New York: Arkana/Penguin, 1992.

Tansley, David. *Dimensions of Radionics: Techniques of Instrumented Distant-Healing*. London: Eastern Press, 1977.

———. *Radionics: A Patient's Guide to Instrumented Distant Diagnosis and Healing*. Dorset, UK: Element, 1985.

———. *The Raiment of Light: A Study of the Human Aura*. London: Routledge and Kegan Paul, 1984.

Thun, Matthias. *The Maria Thun Biodynamic Almanac: North American Edition*. Edinburgh: Floris Books, annual.

Tompkins, Peter. *The Secret Life of Plants: A Fascinating Account of the Physical, Emotional, and Spiritual Relations between Plants*. New York: Harper and Row, 1973.

Cited Works

———. *The Secret Life of Nature: Living in Harmony with the Hidden World of Nature Spirits from Fairies to Quarks.* New York: HarperCollins, 1997.

Tompkins, Peter, and Christopher Bird. *Secrets of the Soil.* New York: Harper and Row, 1989.

Tyson, Donald. *Ritual Magic: What It Is and How to Do It.* St. Paul, MN: Llewellyn, 1992.

Westlake, Aubrey T. *The Pattern of Health: A Search for a Greater Understanding of the Life Force in Health and Disease.* Dorset, UK: Element, 1985.

Wieting, Elizabeth. *Nature Spirits: Endangered like the Bees: How Can We Help Them?* Eugene, OR: New World, 2009.

Willey, Raymond C. *Modern Dowsing: The Dowser's Handbook.* Tucson, AZ: Treasure Chest, 1970.

Wright, Machaelle Small. *The Perelandra Garden Workbook* (ebook ed.). Jeffersonton, VA: Perelandra, 2012.

Yarrow, David. *Dowsing History and Techniques: Beginners Guide.* Auburn, NY: Tri Lake Industries, 1984.

Related Reading

Berg, Peter. *The Moon Gardener: A Biodynamic Guide to Getting the Best from Your Garden*. Edinburgh: Floris Books, 2012.

Bresette-Mills, Jack. *Sensitive Beekeeping: Practicing Vulnerability and Nonviolence with Your Backyard Beehive*. Great Barrington, MA: SteinerBooks, 2016.

Code, Jonathan. *Muck and Mind: Encountering Biodynamic Agriculture: An Alchemical Journey*. Great Barrington, MA: SteinerBooks, 2014.

Cook, Wendy E. *The Biodynamic Food and Cookbook: Real Nutrition that Doesn't Cost the Earth*. Forest Row, UK: Clairview, 2006.

———. *Foodwise: Understanding What We Eat and How It Affects Us: The Story of Human Nutrition*. Forest Row, UK: Clairview, 2003.

Edelglass, Stephen, Georg Maier, Hans Gebert, and John Davy. *The Marriage of Sense and Thought: Imaginative Participation in Science*. Great Barrington, MA: Lindisfarne Books, 2013

Graves, Julia. *The Language of Plants: A Guide to the Doctrine of Signatures*. Great Barrington, MA: Lindisfarne Books, 2012.

Hauk, Gunther. *Toward Saving the Honeybee*. Kimberton, PA: Biodynamic Association, 2017.

Klett, Manfred. *Principles of Biodynamic Spray and Compost Preparations*. Edinburgh: Floris Books, 2005.

Klocek, Dennis. *Sacred Agriculture: The Alchemy of Biodynamics*. Great Barrington, MA: Lindisfarne Books, 2013.

Koepf, Herbert. *The Biodynamic Farm: Agriculture in Service of the Earth and Humanity*. Great Barrington, MA: SteinerBooks, 2006.

König, Karl. *Nutrition from Earth and Cosmos*. Edinburgh: Floris Books, 2015.

———. *Social Farming: Healing Humanity and the Earth*. Edinburgh: Floris Books, 2014.

Kuepper, George. *Radionics, Reality & Man: Experimental Principles and Procedures of Radionics for Personal Health and Well-being*. Goshen, AR: GAIA, 1996.

Related Reading

MacCormack, Harry. *Cosmic Influences on Agricultural Processes*. Corvallis, OR: BioWisdom, 2011.

Massei, Karsten. *School of the Elemental Beings*. Great Barrington, MA: SteinerBooks, 2017.

Morrow, Joel. *Vegetable Gardening for Organic and Biodynamic Growers: Home and Market Gardeners*. Great Barrington, MA: Lindisfarne Books, 2014.

Nastati, Enzo. *Basic Biodynamic Agriculture in 9 Meetings*. Great Barrington, MA: Lindisfarne Books, 2018.

Naydler, Jeremy (ed.). *Goethe on Science: An Anthology of Goethe's Scientific Writings*. Edinburgh: Floris Books, 1996.

Osthaus, Karl-Ernst. *The Biodynamic Farm: Developing a Holistic Organism*. Edinburgh: Floris Books, 2010.

Pfeiffer, Ehrenfried E. *Biodynamic Farming and Gardening Renewal and Preservation of Soil Fertility*. Great Barrington, MA: Portal Books, 2020.

———. *Pfeiffer's Introduction to Biodynamics*. Edinburgh: Floris Books, 2011.

———. *Weeds and What They Tell Us*. Edinburgh: Floris Books, 2012.

Pfeiffer, Ehrenfried E., and Michael Maltas. *The Biodynamic Orchard Book*. Edinburgh: Floris Books, 2013.

Pyle, Jack R., and Taylor Reese. *Raising with the Moon: The Complete Guide to Gardening and Living by the Signs of the Moon*. Asheboro, NC: Down Home Press, 2003.

———. *You and the Man in the Moon: The Complete Guide to Using the Almanac*. Asheboro, NC: Down Home Press, 2003.

Scharff, Paul W. *Commentary on Rudolf Steiner's Agriculture Course: From the Paul W. Scharff Archive*. Great Barrington, MA: SteinerBooks, 2018.

Schwuchow, Jochen, John Wilkes, and Iain Trousdell. *Energizing Water: Flowform Technology and the Power of Nature*. Forest Row, UK: Rudolf Steiner Press, 2010.

Selg, Peter. *The Agriculture Course, Koberwitz, Whitsun 1924: Rudolf Steiner and the Beginnings of Biodynamics*. Forest Row, UK: Rudolf Steiner Press, 2010.

Spock, Marjorie. *Fairy Worlds and Workers: A Natural History of Fairyland*. Great Barrington, MA: SteinerBooks, 2013.

Steiner, Rudolf. *Agriculture: An Introductory Reader*. Forest Row, UK: Rudolf Steiner Press, 2004.

———. *Agriculture Course: The Birth of the Biodynamic Method*. Forest Row, UK: Rudolf Steiner Press, 2004.

———. *Bees*. Hudson, NY: Anthroposophic Press, 1998.

———. *Nature's Open Secret: Introductions to Goethe's Scientific Writings*. Great Barrington, MA: SteinerBooks, 2010.

Strong, Devon. *A Lakota Approach to Biodynamics: Taking Life Seriously*. Great Barrington, MA: Lindisfarne Books, 2016.

Suchantke, Andreas. *Eco-Geography: What We See When We Look at Landscapes*. Great Barrington, MA: Lindisfarne Books, 2001.

Thun, Maria. *Gardening for Life: The Biodynamic Way*. Stroud, UK: Hawthorn Press, 2000.

Thun, Matthias. *Biodynamic Beekeeping: A Sustainable Way to Keep Happy, Healthy Bees*. Edinburgh: Floris Books, 2020.

Waldin, Monty. *Biodynamic Gardening: Grow Healthy Plants and Amazing Produce with the Help of the Moon and Nature's Cycles*. New York: Penguin, 2015.

About the Author

George Kuepper is originally from Wisconsin and has been a resident of the mid-South for many years. He earned a master's degree in science from the University of Wisconsin Department of Agronomy during the mid-1970s and has devoted his life to developing sustainable and organic agriculture for entities such as The Center for the Biology of Natural Systems, St. Louis; The National Center for Appropriate Technology (ATTRA); and Kerr Center for Sustainable Agriculture, Poteau, Oklahoma.

Goerge discovered radionics in the 1980s while struggling to revive an Oklahoma blueberry planting on which agricultural chemicals had been overly applied and misused. The planting was the cornerstone for an eight-acre, U-pick farm project for the Kerr Center. When he learned of radionics, he had reached the point where he was willing to try almost anything.

His first radionics class took place in 1986, after which he returned to the farm with enough confidence to begin turning things around. Diseases and insect pests vanished. The blueberry bushes stopped dying and began to flourish. Even the weeds backed off from their earlier onslaught. The transformation was profound.

Since that time, George has written many training manuals explaining the foundations of radionics, the use and maintenance of instruments, fundamental operations, and specific applications. Those books were intended to supplement classroom and training. George also offers training for individuals and small-groups, primarily for beginners, and sells instruments for implementing the practices of radionics and dowsing.

Visit George Kuepper's website at www.midsouthradionics.com.

www.ingramcontent.com/pod-product-compliance
Lightning Source LLC
Chambersburg PA
CBHW051123160426
43195CB00014B/2318